Balazs Pritz

**LES of the Pulsating, Non-Reacting Flow in Combustion Chambers**

Balazs Pritz

# LES of the Pulsating, Non-Reacting Flow in Combustion Chambers

## A Numerical Investigation for the Prediction of the Stability Limits of Combustion Systems

Südwestdeutscher Verlag für Hochschulschriften

## Impressum/Imprint (nur für Deutschland/ only for Germany)

Bibliografische Information der Deutschen Nationalbibliothek: Die Deutsche Nationalbibliothek verzeichnet diese Publikation in der Deutschen Nationalbibliografie; detaillierte bibliografische Daten sind im Internet über http://dnb.d-nb.de abrufbar.

Alle in diesem Buch genannten Marken und Produktnamen unterliegen warenzeichen-, marken- oder patentrechtlichem Schutz bzw. sind Warenzeichen oder eingetragene Warenzeichen der jeweiligen Inhaber. Die Wiedergabe von Marken, Produktnamen, Gebrauchsnamen, Handelsnamen, Warenbezeichnungen u.s.w. in diesem Werk berechtigt auch ohne besondere Kennzeichnung nicht zu der Annahme, dass solche Namen im Sinne der Warenzeichen- und Markenschutzgesetzgebung als frei zu betrachten wären und daher von jedermann benutzt werden dürften.

Verlag: Südwestdeutscher Verlag für Hochschulschriften Aktiengesellschaft & Co. KG
Dudweiler Landstr. 99, 66123 Saarbrücken, Deutschland
Telefon +49 681 37 20 271-1, Telefax +49 681 37 20 271-0
Email: info@svh-verlag.de
Zugl.: Karlsruhe, KIT, Diss., 2009

Herstellung in Deutschland:
Schaltungsdienst Lange o.H.G., Berlin
Books on Demand GmbH, Norderstedt
Reha GmbH, Saarbrücken
Amazon Distribution GmbH, Leipzig
ISBN: 978-3-8381-1304-3

## Imprint (only for USA, GB)

Bibliographic information published by the Deutsche Nationalbibliothek: The Deutsche Nationalbibliothek lists this publication in the Deutsche Nationalbibliografie; detailed bibliographic data are available in the Internet at http://dnb.d-nb.de.

Any brand names and product names mentioned in this book are subject to trademark, brand or patent protection and are trademarks or registered trademarks of their respective holders. The use of brand names, product names, common names, trade names, product descriptions etc. even without a particular marking in this works is in no way to be construed to mean that such names may be regarded as unrestricted in respect of trademark and brand protection legislation and could thus be used by anyone.

Publisher: Südwestdeutscher Verlag für Hochschulschriften Aktiengesellschaft & Co. KG
Dudweiler Landstr. 99, 66123 Saarbrücken, Germany
Phone +49 681 37 20 271-1, Fax +49 681 37 20 271-0
Email: info@svh-verlag.de

Printed in the U.S.A.
Printed in the U.K. by (see last page)
ISBN: 978-3-8381-1304-3

Copyright © 2010 by the author and Südwestdeutscher Verlag für Hochschulschriften Aktiengesellschaft & Co. KG and licensors
All rights reserved. Saarbrücken 2010

# Contents

Abstract ............................................................................................................. III
Zusammenfassung ............................................................................................ IV
Acknowledgements .......................................................................................... VI
Nomenclature .................................................................................................. VII
Papers ............................................................................................................. XII
1 Introduction ................................................................................................. 1
   1.1  Motivation ............................................................................................ 1
   1.2  Object of the Thesis ............................................................................. 2
2 Combustion Instabilities ............................................................................. 4
        Suppression of Combustion-Driven Oscillations ................................ 5
        System Analysis ................................................................................... 6
        Helmholtz Resonator ........................................................................... 7
        The Reduced Physical Model .............................................................. 8
3 Methodology .............................................................................................. 11
   3.1  Governing Equations ......................................................................... 12
   3.2  Turbulence Modelling ....................................................................... 13
        Large Eddy Simulation ...................................................................... 15
   3.3  Full Multigrid Method ....................................................................... 17
   3.4  Dual Time Stepping ........................................................................... 17
   3.5  Mesh generation ................................................................................. 18
   3.6  Boundary Conditions ......................................................................... 19
      3.6.1  Wall ......................................................................................... 19
        Near-Wall Treatment of Turbulence ................................................ 19
        Wall Roughness – The Discrete Element Method ........................... 21
      3.6.2  Non-Reflecting Boundary Condition .................................... 23
      3.6.3  Mass Flow Rate Inlet ............................................................. 25
      3.6.4  Precursor Type Inlet Condition ............................................. 26
      3.6.5  Synthetic Eddy Method ......................................................... 33

| 4 | Single Resonator: Combustion Chamber | 38 |
|---|---|---|
| 4.1 | Computational Domain and Boundary Conditions | 40 |
| 4.2 | Flow Properties in the Combustion Chamber | 42 |
| 4.3 | Damping Mechanisms | 45 |
| 4.4 | Comparison of the Results with Experimental Data | 52 |
| 4.4.1 | Case II: Basic Configuration | 52 |
| 4.4.2 | Case III: Variation of the Geometry | 54 |
| 4.4.3 | Case IV: Variation of the Fluid Temperature | 55 |
| 4.4.4 | Case V: Simulation with Roughness | 57 |
| 5 | Coupled Resonators: Burner and Combustion Chamber | 61 |
| 5.1 | Computational Domain and Boundary Conditions | 63 |
| 5.2 | Flow Properties in the Coupled System | 66 |
| 5.3 | Comparison of the Results with Experimental Data | 67 |
| 6 | On the Source of Combustion Instabilities | 70 |
| 7 | Computational Effort | 74 |
| 8 | Conclusions and Outlook | 75 |
| References | | 78 |
| List of Tables | | 92 |
| List of Figures | | 93 |

## Abstract

It is well known that in order to fulfil the stringent demands for low emissions of $NO_x$, the lean premixed combustion concept is commonly used. However, lean premixed combustors are susceptible to thermo-acoustic instabilities driven by the combustion process and possibly sustained by a resonant feedback mechanism coupling pressure and heat release. This resonant feedback mechanism creates pulsations typically in the frequency range of several hundred *Hz*, which reach high amplitudes so that the system has to be shut down or it is even damaged. Although the research activities of the recent years have contributed to a better understanding of this phenomenon, the underlying mechanisms are still not understood well enough.

For the prediction of the stability of technical combustion systems regarding the development and maintaining of self-sustained combustion instabilities the knowledge of the periodic-non-stationary mixing and reacting behaviour of the applied flame type and a quantitative description of the resonance characteristics of the gas volumes in the combustion chamber is conclusively needed. In the present work the numerical investigation of the pulsating, non-reacting flow in a combustion chamber, and a coupled system of burner plenum and combustion chamber is presented. The results are compared with an analytical model and experimental data.

By means of the numerical simulation and the analytical model the resonance characteristics of a combustion system can be predicted already in the design stage. The main goal of the numerical investigation is to predict the damping coefficient of the system, which is an important input for the reduced physical model. In order to provide a detailed insight into the flow mechanics inside the system Large Eddy Simulations (LES) were carried out.

The numerical simulations enabled to analyse several phenomena of the pulsating flow and they provided for the identification of the main damping mechanisms.

The results of the investigation of the single resonator and the coupled resonators showed that LES can predict accurately the resonant characteristics of the system and the damping, respectively.

## Zusammenfassung

Für die erfolgreiche Umsetzung fortschrittlicher Verbrennungskonzepte in Turbinenbrennkammern und Industriefeuerungen ist die Vermeidung selbsterregter Verbrennungsinstabilitäten eine unverzichtbare Voraussetzung. Die grundlegende Untersuchung dieser Phänomene ist daher eine wesentliche Zielsetzung des Sonderforschungsbereiches 606 (SFB 606): "Instationäre Verbrennung: Transportphänomene, Chemische Reaktionen, Technische Systeme" an der Universität Karlsruhe (TH). Um die Schwingungsstabilität eines Verbrennungssystems, das im einfachsten Fall aus den Komponenten Brenner-Flamme-Brennkammer besteht, bereits bei der Konzeption in Abhängigkeit anwendungstechnisch relevanter Betriebsparameter (gewünschte Last- und Luftzahlregelbereiche, Brennstoffart, Gasqualität, Luftvorwärmung) vorhersagen zu können, ist es erforderlich, das von der Anregungsfrequenz und -amplitude sowie von den voran genannten Betriebsparametern abhängige Übertragungsverhalten aller Komponenten im Rückkopplungskreis quantitativ beschreiben zu können.

Hierzu wurde das frequenzabhängige Druckübertragungsverhalten einer Brennkammer vom Helmholtz-Resonator-Typ modelliert und mit experimentell bestimmten Resonanzkurven verglichen. In diesen Untersuchungen wurde in der Brennkammer keine Flamme betrieben, das System wurde zwangserregt. Im Rahmen der Untersuchungen wurden auch numerische Simulationen durchgeführt, die in der vorliegenden Arbeit dargestellt werden. Die Strömungsmodellierung wurde mit Hilfe der Grobstruktursimulation (Large Eddy Simulation - LES) für die gesamte Brennkammerdurchströmung durchgeführt.

Um das Schwingungsverhalten der Rauchgassäule in einer realen, gedämpften Brennkammer vom Helmholtz-Resonator-Typ frequenzabhängig und für verschiedene Geometrien vorhersagen zu können, ist es erforderlich, die verschiedenen Mechanismen, die zu einer Dissipation der Schwingungsenergie und somit zu einer endlichen Begrenzung der auftretenden Druckamplituden im Resonanzfall führen, zu identifizieren und hinsichtlich ihrer Bedeutung für anwendungstechnisch relevante Brennkammerkonfigurationen zu beurteilen. Bei Kenntnis des für die jeweilige Brennkammergeometrie maßgebenden Dämpfungsmechanismus ist es dann möglich, das integrale Dämpfungsmaß, welches den frequenzabhängigen Verlauf der Resonanzkurve maßgeblich festlegt, zu berechnen und in ein Schwingungsmodell zu integrieren. Zielsetzung des

Teilprojektes A7 war in einem ersten Schritt die experimentelle Bestimmung des integralen Dämpfungsmaßes einer Modellbrennkammer als einfacher Resonator. Hierbei wurden die charakteristischen Abmessungen der Brennkammer, die die Resonanzfrequenz und die Schwingungsdämpfung bestimmen, sowie die Temperatur des zwangserregten Luftmassenstromes systematisch variiert.

Die Ergebnisse der numerischen Untersuchung zeigten eine sehr gute Übereinstimmung mit dem analytischen Modell und mit dem experimentell ermittelten Resonanzverhalten der Brennkammer. Mit Hilfe der detailreichen Wiedergabe der Strömung durch numerische Simulation wurden Ort und Mechanismen der Schwingungsdämpfung analysiert.

Die nachfolgende Aufgabe war, das gekoppelte Schwingungsverhalten einer Brennkammer mit vorgeschaltetem Gasvolumen, in diesem Fall dem Brennergehäuse, zu beschreiben. Dieses Vorgehen war erforderlich, da bereits geringe Gasvolumina stromab oder stromauf der Brennkammer zu erheblichen Abweichungen des gekoppelten Resonanzverhaltens gegenüber dem einfachen Helmholtz-Resonator führen. Die numerische Simulation lieferte in diesem Fall auch eine gute Übereinstimmung mit den experimentellen Daten.

Es ist damit möglich mit Hilfe von LES und dem analytischen Modell die Vorhersage der Stabilitätsgrenzen eines Verbrennungssystems, welches aus Komponenten des Helmholtz-Resonator-Typs besteht, in der Konzeptionsphase zu ermitteln. Die zukünftige Aufgabe besteht darin, die Wärmefreisetzung der Flamme zu berücksichtigen. Letztendlich ist es gewünscht, die Simulation des Systems einschließlich pulsierender Flamme durchführen zu können.

## Acknowledgements

This thesis was created during my PhD study from November 2002 until Mai 2009 at the Department of Fluid Machinery, Faculty of Mechanical Engineering at the University of Karlsruhe (TH). My work was founded by the Karlsruher Universitätsgesellschaft, by the German Academic Exchange Service (DAAD) and by the German Research Foundation (DFG) through the Project A7 in the Collaborative Research Centre (SFB) 606.

The calculations have been performed at the SSC Karlsruhe, HLRS of the University of Stuttgart and LRZ Munich.

I would like to express special thanks to Prof. Martin Gabi, who gave me the opportunity to join to his CFD research group and examined my PhD thesis. I also wish to thank Dr. Franco Magagnato, the leader of this CFD group, for his experienced advice and valuable explanations about numerics, turbulence and Large Eddy Simulation.

I am particularly grateful to Prof. Horst Büchner for his support and for agreeing to be co-examiner for my PhD thesis. I also like to thank him and his PhD students for explaining me the experimental setup and providing me experimental data.

I am indebted to Dr. László Kullmann (Budapest University of Technology and Economics) for his great support and for agreeing to be co-examiner for my PhD thesis.

My time in the Department of Fluid Machinery would not have been nearly as pleasant without my colleagues. I am gratefully for their friendship, cooperation, help and interest in my work. I would like to thank them all very much.

I would also like to thank my family for their support, love and encouragement. I especially thank to my wife, Kriszta.

Karlsruhe, 02-07-2009

# Nomenclature

**Symbols**

| | | |
|---|---|---|
| $A$ | [-] | Amplitude ratio of mass flow rates (Eq. (4.4)) |
| $A$ | [$m^2$] | Area |
| $a$ | [-] | Amplitude ratio for pulsating flows |
| $B$ | [$kg/s$] | Damping coefficient |
| $C_S$ | [-] | Smagorinsky constant |
| $c$ | [$m/s$] | Speed of sound |
| $c_f$ | [-] | Drag coefficient |
| $c_p$ | [$kJ/kg\,K$] | Specific heat at constant pressure |
| $c_v$ | [$kJ/kg\,K$] | Specific heat at constant volume |
| $D$ | [-] | Damping factor |
| $d$ | [$m$] | Diameter |
| $F$ | [$N$] | Force |
| $f$ | [$N/m^3$] | Specific force |
| $f$ | [$Hz$] | Frequency |
| $f_0$ | [$Hz$] | Eigenfrequency (resonant frequency) |
| $g$ | [$m/s^2$] | Body force |
| $h$ | [$m$] | Height; half channel height |
| $i, j$ | [-] | Run index |
| $k$ | [$N/m$] | Spring stiffness |
| $k$ | [$m$] | Roughness height |
| $k$ | [$m^2/s^2$] | Turbulent kinetic energy per unit mass |
| $L, l$ | [$m$] | Length |
| $M$ | [$m$] | Grid size of the turbulence generator grid |
| $m$ | [$kg$] | Mass |
| $\dot{m}$ | [$kg/s$] | Mass flow rate |

| | | |
|---|---|---|
| $Ma$ | [-] | Mach number |
| $\vec{n}$ | [-] | Surface normal |
| $n$ | [-] | Run index |
| $p$ | [$Pa$] | Pressure |
| $Q$ | [$s^{-2}$] | $Q$-criterion (Eq. (3.34)) |
| $\dot{q}$ | [$W/m^2$] | Heat release rate |
| $R$ | [$kJ/kg\,K$] | Specific gas constant |
| $R$ | [$m$] | Radius |
| $r$ | [$m$] | Radial coordinate |
| $R_{ij}$ | [$m^2/s^2$] | Correlation tensor |
| $Re$ | [-] | Reynolds number |
| $S$ | [$m^2$] | Surface |
| $S_{ij}$ | [$s^{-1}$] | Strain rate tensor |
| $St$ | [-] | Strouhal number |
| $T$ | [$K$] | Temperature |
| $T$ | [$s$] | Period of pulsation |
| $\mathbf{T}$ | [$N/m^2$] | Viscous shear stress tensor |
| $Tu$ | [%] | Turbulence intensity |
| $t$ | [$s$] | Time |
| $U$ | [$m/s$] | Characteristic velocity |
| $u, v, w$ | [$m/s$] | Velocity components |
| $\mathcal{u}$ | [-] | Normalized signal for the fluctuating part of the velocity for SEM |
| $V$ | [$m^3$] | Volume |
| $v$ | [$m/s$] | Velocity |
| $x, y, z$ | [$m$] | Coordinates |

**Greek symbols**

| | | |
|---|---|---|
| $\alpha$ | [°] | Angle |

| | | |
|---|---|---|
| $\gamma$ | [-] | Specific heat ratio |
| $\Delta$ | [m] | Filter length scale |
| $\delta$ | [m] | Boundary layer thickness |
| $\delta_{ij}$ | [-] | Kronecker symbol |
| $\varepsilon$ | [$m^2/s^3$] | Dissipation of turbulent kinetic energy |
| $\kappa$ | [$m^{-1}$] | Wave number |
| $\mu$ | [Pa s] | Dynamic viscosity |
| $\nu$ | [$m^2/s$] | Kinematic viscosity |
| $\rho$ | [$kg/m^3$] | Density |
| $\tau$ | [s] | Time scale for turbulent kinetic energy |
| $\tau$ | [$N/m^2$] | Shear stress |
| $\Phi$ | [-] | Generic variable |
| $\Phi$ | [$J/s\ m^3$] | Dissipation function |
| $\varphi$ | [°] | Phase shift angle |
| $\Omega_{ij}$ | [$s^{-1}$] | Vorticity tensor |
| $\omega$ | [$s^{-1}$] | Frequency of dissipation rate of the turbulent kinetic energy |
| $\omega$ | [$s^{-1}$] | Angular frequency |

**Subscripts**

| | |
|---|---|
| bp | Burner plenum |
| c | Cone |
| cc | Combustion chamber |
| egp | Exhaust gas pipe |
| ex | Excitation |
| fl | Fluid |
| in | Inlet |
| mfr | Mass flow rate |
| out | Outlet |

| | |
|---|---|
| puls | Pulsation |
| ref | Reference |
| rms | Root mean square |
| r | Roughness |
| s | Sand |
| ss | Supply system |
| st | Steady |
| uc | Centreline velocity |
| t | Turbulent; tangential |
| W | Wall |

## Superscripts

| | |
|---|---|
| $\overline{\Phi}$ | Mean; resolved (filtered) part |
| $\hat{\Phi}$ | Amplitude |
| $\tilde{\Phi}$ | Pulsating component |
| $\Phi'$ | Fluctuating component |
| $\Phi^+$ | Normalized value |
| $\vec{\Phi}$ | Vector |

## Abbreviations

| | |
|---|---|
| BC | Boundary condition |
| CDS | Central differencing scheme |
| CFD | Computational Fluid Dynamics |
| CFL | Courant-Friedrichs-Lewy number |
| CRC | Collaborative Research Centre, (see SFB) |
| CV | Control volume |
| DEM | Discrete Element Method |
| DNS | Direct Numerical Simulation |

| | |
|---|---|
| EF | Eigenfrequency (resonant frequency) |
| FMG | Full Multigrid |
| FVM | Finite Volume Method |
| HLRS | Höchstleistungsrechenzentrum der Universität Stuttgart (High Performance Computing Centre Stuttgart) |
| HPF | High-pass filtered |
| HR | Helmholtz resonator |
| LES | Large Eddy Simulation |
| LP | Lean premixed |
| LRZ | Leibniz-Rechenzentrum, computer centre in Munich |
| mfr | Mass flow rate |
| MPI | Message Passing Interface |
| PIV | Particle Image Velocimetry |
| RANS | Reynolds Averaged Navier-Stokes |
| RMS | Root mean square |
| SEM | Synthetic Eddy Method |
| SFB | Sonderforschungsbereich (see CRC) |
| SGS | Subgrid-scale |
| SP | Subproject |
| SPARC | Structured Parallel Research Code |
| SSC | Scientific Supercomputing Centre |
| TAO | Thermo-acoustic oscillation, thermal acoustic oscillation |
| TKE | Turbulent kinetic energy |
| UDS | Upwind differencing scheme |
| URANS | Unsteady RANS |

# Papers

The content of this thesis is based on the following publications which have been published or have been submitted for publications during the course of the PhD project.

Magagnato, F., Pritz, B., Büchner, H., Gabi, M.: Prediction of the Resonance Characteristics of Combustion Chambers on the Basis of Large-Eddy Simulation. *Journal of Thermal Science*, **14** (2), pp.156-161, 2005.

Petsch, O., Pritz, B., Magagnato, F., Büchner, H.: Untersuchungen zum Resonanzverhalten einer Modellbrennkammer vom Helmholtz-Resonator-Typ. In *Verbrennung und Feuerungen - 22. Deutscher Flammentag, VDI-GET, VDI-Berichte* Nr. 1888, 507-512, Braunschweig, 21-22. September, 2005; ISBN 3-18-091888-8

Magagnato, F., Pritz, B., Gabi, M.: Calculation of a Turbine Blade at High Reynolds Numbers by Large-Eddy Simulation. *Proceedings of the 11th International Symposium on Transport Phenomena and Dynamics of Rotating Machinery*, Honolulu, Hawaii, February 26 - March 2, 2006, paper 87 (CD-ROM)

Pritz, B., Magagnato, F., Gabi, M.: Inlet Condition For Large Eddy Simulation Applied To a Combustion Chamber. *Proceedings of the 13th International Conference on Fluid Flow Technologies*, Budapest, Hungary, September 6-9, 2006; ISBN 963 420 872 X

Magagnato, F., Pritz, B., Gabi, M.: Inflow Conditions for Large-Eddy Simulation of Compressible Flow in a Combustion Chamber. *Proceedings of the 5th International Symposium on Turbulence, Heat and Mass Transfer*, Dubrovnik, Croatia, September 25-29, 2006; ISBN 1-56700-229-3

Magagnato, F., Pritz, B., Gabi, M.: Generation of Inflow Conditions for Large-Eddy Simulation of Compressible Flows. *Proceedings of the 8th International Symposium on Experimental and Computational Aerothermodynamics of Internal Flows (ISAIF)*, Lyon, France, July 2-5, 2007

Magagnato, F., Pritz, B., Gabi, M.: Calculation of the VKI Turbine Blade with LES und DES. *Journal of Thermal Science*, **16** (4), pp. 321-327, 2007.

Pritz, B., Magagnato, F., Gabi, M.: Investigation of the Effect of Surface Roughness on the Pulsating Flow in Combustion Chambers with LES. *Proceedings of the EU-Korea Conference on Science and Technology*, Heidelberg, Germany, 2008.

Magagnato, F., Pritz, B., Gabi, M.: A Novel Approach to Predict the Stability Limits of Combustion Chambers with Large Eddy Simulation. *Proceedings of the 9th International Symposium on Experimental and Computational Aerothermodynamics of Internal Flows (ISAIF)*, Gyeongju, Korea, September 8-11, 2009.

Pritz, B., Magagnato, F., Gabi, M.: Stability Analysis of Combustion Systems by Means of Large Eddy Simulation. *Proceedings of the 14th International Conference on Fluid Flow Technologies*, Budapest, Hungary, September 9-12, 2009.

# 1 Introduction

## 1.1 Motivation

Our primary energy consumption is supported in 81% by the combustion of fossil energy commodities [50]. The demand on energy will grow by about 60% in the near future [126]. The efficiency of the combustion processes is crucial for the environment and for the use of the remaining resources. At the University of Karlsruhe the long-term project Collaborative Research Centre (CRC) 606: "Non-stationary Combustion: Transport Phenomena, Chemical Reactions, Technical Systems" is founded to investigate the basics of combustion and for the implementations relevant processes coupled to combustion [9], [148].

Modern combustion concepts comprise lean premixed (LP) combustion, which allows for the reduction of the pollutant emissions, in particular oxides of nitrogen ($NO_x$) [64]. Lean premixed combustors are, however, prone to combustion instabilities with both low and high frequencies. These instabilities result in higher emission, acoustical load of the environment and even in structural damage of the system.

The subproject (SP) A7 in CRC 606 is dedicated to investigate low frequency instabilities in combustion systems. The main goal of the SP A7 is to validate an analytical model which was developed by Büchner [14] to describe the resonant characteristics of combustion systems. Further goal of SP A7 was to build scaling laws for geometry and operation condition parameters (mass flow rate, temperature). The ultimate task of the analytical model is to predict the stability limits of a system already in the design stage. A very important input for the physical model is the damping factor $D$ [14]. The damping can be measured experimentally or predicted by numerical simulation. In order to use the physical model in the design stage a reliable numerical tool is needed. In the SP A7 numerical simulations were carried out to prove the abilities of the simulation to predict the damping factor, furthermore to localize the effects mostly responsible for the damping and to gain a detailed insight into the pulsating flow. A successful prediction of the damping factor in dependence of the resonator geometry and the fluid temperature, as well, has not been carried out by numerical simulation before.

The present work renders the numerical investigations of the SP A7.

## 1.2 Object of the Thesis

It is an indispensable prerequisite for the successful implementation of advanced combustion concepts to avoid periodic combustion instabilities in combustion chambers of turbines and in industrial combustors [15], [63]. For the elimination of the undesirable oscillations it is important to know the mechanisms of feedback of periodic perturbations in the combustion system. If the transfer characteristics of the subsystems (in a simple case burner, flame and chamber) furthermore of the coupled subsystems are known, the oscillation disposition of the combustion system can be evaluated during the design phase for different, realistic operation conditions (desired load range, air ratio, fuel type, fuel quality and temperature).

The investigations of the low-frequency oscillations in the range of a few $Hz$ up to several 100 $Hz$ were focused on the passive parts of the system: the combustion chamber and the burner plenum. The determination of the flame resonant characteristics is the object of other works [14], [31], [73], [74], [75], and also of the SP C1 within the CRC 606.

The investigations of SP A7 were arranged in two phases. In the first phase a Helmholtz-resonator type combustion chamber was investigated. In the second phase the physical model was extended for a coupled system of burner plenum and combustion chamber. The resonant characteristics of the system were determined by the usual method used for stability analysis. The system was excited at the inlet with a well defined sinusoidal signal and the response of the system was detected at the outlet. A sinusoidal varying mass flow rate was prescribed at the inlet and the mass flow rate at the outlet cross section of the exhaust gas pipe was registered.

In the present investigations the flow is non-reacting. There is no combustion, thus no flame in the combustion chamber. Hence there is no self-excited thermo-acoustic oscillation, the pulsation is forced externally. In the experiment there was a pulsator unit, which produced a sinusoidal component of the mass flow rate [14].

Parallel to the experimental measurements numerical simulations were carried out to identify the resonant characteristics of the combustion system. An attempt was made previously to determine the damping factor $D$ by means of unsteady Reynolds-averaged Navier-Stokes simulation (URANS), but the results were not satisfactory [115]. In order to understand better the flow effects in the combustor and to localize the main dissipation Large Eddy Simulation (LES) was chosen.

In the first section of the thesis the state of the art in terms of combustion instabilities is reviewed. Next the used methodology is depicted and the main functionality of the in-house solver SPARC is shown. In the second half of the thesis the investigation of the single resonator is discussed beginning with the definition of the geometry and ending with the comparison of the results with experimental data. The flow in the exhaust gas pipe is discussed in more detail. The case of the coupled resonators is discussed subsequently. In the last section a source of the combustion instability and a novel approach is described. Latter can provide the eigenfrequency of the system if the use of the Helmholtz theory is inapplicable because of e.g. the complexity of the geometry. At least a short outlook is taken with recommendations for further investigations.

## 2 Combustion Instabilities

First it is important to clarify that within the frame of this work the ignition stability of the flame will not be concerned. The combustion instabilities described here are driven by thermo-acoustic self-excited oscillations. If there is no pulsation in the combustion chamber the flame is stable. Furthermore pulse combustors designed for oscillations are also not dealt within this work [111], [152].

In order to get a high density of heat release flux i.e. power density and simultaneously low $NO_x$ emission highly turbulent lean premixed or partially premixed flames are mostly used [64]. Significant property of these flames is that any disturbances in the equivalence ratio through turbulence or in the air/fuel mixture supply produce a very fast change in the heat release. Compared to axial jet flames the premixed swirl flames can significantly amplify the disturbances [16]. The combustion process is increasingly sensitive to perturbation in the equivalence ratio under lean operating conditions.

Unsteady heat release involves pressure and velocity pulsation in the combustion chamber. These can result in thrust oscillation in aircraft engines, enhanced heat transfer and thermal stresses to combustor walls and other system components, oscillatory mechanical loads that results in low- and high-cycle fatigue of system components [58], [69]. The oscillation of flow parameters can increase the amplitude of flame movements. This can cause blowoff of the flame or, in worst case, a flashback of the flame into the burner plenum. There are several mechanisms suspected of leading to combustion instabilities, such as periodic inhomogeneities in the mixture fraction, pressure sensitivity of the flame speed and the formation of large-scale turbulent structures.

The coupling of flame and acoustics can produce self-excited thermo-acoustic pulsation. The pulsation will be amplified then to the "limit cycle". Thermo-acoustic or thermal acoustic oscillations (TAO) were observed at first by Higgins in 1777 during his investigation of a "singing flame" [39]. The computation of self-excited thermo-acoustic oscillations began with the investigation of the Rijke-tube by Lehmann in 1937 [65]. A short overview about the history of simulations of TAO is given in [37]. It shows that most of the investigators wanted to compute oscillations excited by the flame or the system with flames excited by an external force at least. Because of the complexity of the problem many computations could not predict the limit cycle.

Lord Rayleigh proposed for the first time a criterion, which, regardless of the source of the instabilities, describes the necessary condition for instabilities to occur [109]. The criterion expresses that a pressure oscillation is amplified if heat is added at a point of maximum amplitude or extracted at a point of minimum amplitude. If the opposite occurs, a pressure oscillation is damped. The mathematical representation of this criterion was first proposed by Putnam and Dennis [107] as:

$$\int_0^T \tilde{q}(t) \cdot \tilde{p}(t) dt > 0 \qquad (2.1)$$

where $\tilde{q}$ and $\tilde{p}$ are the fluctuating parts of the heat release rate and the pressure, respectively. The condition will be satisfied for a given frequency if the phase difference between the heat release oscillation and the pressure oscillation is less than ±90°. Additionally, the amplitude of the pressure oscillation will be amplified if the losses through the damping effects are less than the energy fed into the oscillation. More appropriate forms of the Rayleigh criterion and similar criterions can be found in [104].

**Suppression of Combustion-Driven Oscillations**

In combustion systems of highly complex shape there can be more various modes: low frequency bulk mode, transversal, tangential, radial and longitudinal modes. In such a combustion system it is almost impossible today to predict all the unstable operating points. There are more strategies in practice to suppress the combustion oscillations in the unstable operating points. These can be grouped into passive and active control methods.

Passive or static control methods tune the resonance characteristics of the combustion system with additional devices as quarter-wave tube, Helmholtz resonators, sound-absorbing batting, orifice, ports and baffles [107]. Resonators can be placed in the fuel system [113] or in the combustor [36] or in other component. Perforates can be used at the premixer inlet [140], which is also an additional resonator to tune the resonant characteristics of the system. Instabilities can also be suppressed by means of injection of aluminium [38]. Passive or static control strategies methods are more robust and need a minimum of maintenance. Their disadvantage is that while an unstable operating point is removed, another may arise.

An overview about theory and practice of active control methods is given in [2]. Active control methods can be subdivided into open-loop [112] and closed-loop design [61]. Active control is achieved by a sensor in the combustion chamber, which measures frequency and phase of the combustion oscillation. The measured signal is analyzed then and a proper periodic response is determined. The response is either an acoustic perturbation [119] or a modulation of the fuel injection [35]. Active control is able to suppress combustion instabilities substantially and is already in use for numerous practical applications. However, the apparatus is rather expensive and needs continuous maintenance. A failure of the control system can lead to a break down of the combustion system.

Based on the investigations of combustion instabilities [19], [153] there is also an approach to keep off unstable regimes during altering operation conditions. Online prediction of the onset of the combustion instabilities can help the operator to avoid, that the system becoming unstable [57], [68], [149]. This technique is very useful if the ambient conditions vary in wide range e.g. for aircraft gas turbine. For stationary gas turbines with approximately constant ambient conditions, however, this cannot help to design the system for operation conditions, where combustion instabilities are not present.

**System Analysis**

In order to analyse the stability of the system control theory can be used. The combustion system can be divided in subsystems as burner plenum, flame and combustion chamber [6], [14], [67], [105]. The simplified feedback loop of these subsystems is depicted in Figure 2.1. A perturbation of the pressure in the combustion chamber influences the mass flow rate at the burner outlet. This changes the heat release rate of the flame, which results in an alteration of the pressure in the combustion chamber. The transfer function of this closed loop and the subsystems can be determined by system identification furthermore the stability can be investigated by e.g. the Nyquist criterion [20], [120], [121].

If the system is built from these elements, a thermoacoustic network can be modelled to predict the unstable modes [8]. Here, however, some information from measurement is needed.

If the phase shift and gain of the components is known the amplification of the pulsation can be predicted by means of the Rayleigh criterion. This shows that the accurate knowledge of the phase and gain relationship between pressure and heat release oscillation is a key issue to design stable combustion systems.

# Combustion Instabilities

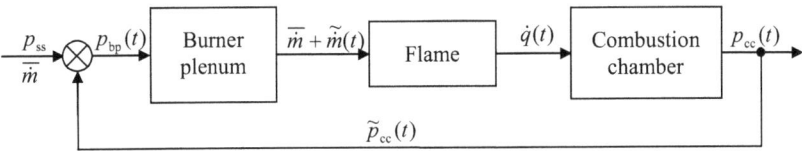

**Figure 2.1:** Feedback loop of a combustion system

## Helmholtz Resonator

Helmholtz resonators are mostly used as passive devices for attenuations of pulsations in combustion systems. Furthermore the resonance behaviour of the combustion system can be described if it bears analogy to this resonator.

If a cavity is coupled to the ambient through a port (Figure 2.2), the gas in this system can be forced into resonance if excited with a certain frequency. Such a geometrical configuration is named Helmholtz resonator after Hermann von Helmholtz, who investigated such devices in the 1850s. The port is the resonator neck, the cavity is the resonator.

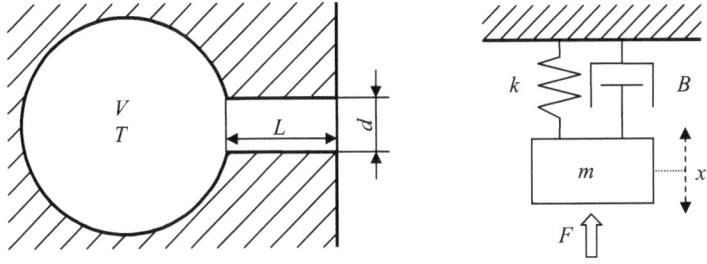

**Figure 2.2:** The Helmholtz resonator and a mass-spring-damper system

The mechanical counterpart of the Helmholtz resonator is a mass-spring-damper system (Figure 2.2). The gas in the neck acts as the mass, the gas in the cavity acts as the spring. The identification of the damping is more difficult. There are linear and non-linear effects in the flow. Damping is provided by the bulk viscosity during the pressure-volume work, the laminar viscosity in the oscillating boundary layer in the resonator neck, the vortex shedding at the ends of the resonator neck at the inflow and outflow and the dissipation of the kinetic energy through turbulence generation. It will be discussed in Section 4.3 which source is dominating in the pulsating flow in the combustion system.

The eigenfrequency of the Helmholtz resonator can be predicted as:

$$f_0 = \frac{c}{2 \cdot \pi} \cdot \sqrt{\frac{A}{V \cdot \left(L + \frac{\pi}{4} \cdot d\right)}} \qquad (2.2)$$

where $c$ is the speed of sound and can be calculated from the temperature $T$ of an ideal gas as:

$$c = \sqrt{\gamma R T}. \qquad (2.3)$$

Furthermore in Equation (2.2) $d$ is the diameter, $L$ is the length and $A$ is the cross section area of the neck, $V$ is the volume of the resonator. The second term in the parenthesis in the denominator is a length correction term, which can be different for Helmholtz resonators with different geometries.

In order to describe the resonance behaviour of combustion systems, they can be treated as single or coupled Helmholtz resonators. In combustion systems the combustion chamber, the burner plenum or other components with larger volume act as resonators. The exhaust gas pipe and the components coupling the resonator volumes together are resonator necks. In industrial combustors the identification of the components of the Helmholtz resonators is easier, in gas turbines more difficult. It is very important which components are assumed to be coupled and which are decoupled. Wrong assumptions can lead to predicting modes incorrectly or even it is impossible to predict certain modes.

**The Reduced Physical Model**

The suppression of the combustion oscillations is not a universal solution. The main goal is to design the combustion system not to be prone to combustion instabilities.

For the prediction of the stability of combustion systems regarding the development and maintaining of self-sustained combustion instabilities the knowledge of the periodic-non-stationary mixing and reacting behaviour of the applied flame type and a quantitative description of the resonance characteristics of the gas volumes in the combustion chamber is conclusively needed. In order to describe the periodic combustion instabilities many attempt have been made to assign the dominant frequency of oscillation to the geometry of the combustion chamber. For the description of the geometry-dependent resonance frequency of the system the equations were derived under the assumption of undamped oscillation (e.g. ¼ wave resonator, Helmholtz resonator). These models predict the resonance frequency quite accurate since the shift due to the moderate damping in the system is negligible. Such a simplified model, however, is not applicable for a quantitative prediction of the stability limit of a real combustion system. On one hand it predicts infinite amplification at the resonance frequency. On the other hand the frequency-dependent phase shift between input and output is described by a step function, hence it cannot be used for the application of a phase criterion (Rayleigh or Nyquist criterion), which is used to predict the occurrence of pressure and heat release oscillations in real combustion systems.

A reduced physical model was developed by Büchner [14], which is able to describe the resonance characteristics of combustion chambers, if their geometry satisfies the geometrical conditions of a Helmholtz resonator [4], [14], [72], [103], [116]. The reduced physical model was derived similar to the resonance behaviour of a mass-spring-damper system, which provides a continuous transfer function of the amplification and the phase shift. First the model was developed to describe a single resonator, later it was extended to a coupled system of two resonators. For this reduced physical model scaling laws were developed based on experimental data. The influence of the amplitude of pulsation, the mean mass flow rate, the temperature of the gas and the geometry were investigated.

In this model the damping in the system is expressed by an integral value. The damping factor cannot be determined by analytical solution. The accurate determination of the damping based on the $2^{nd}$ Rayleigh-Stokes problem is not possible because of the complexity and non-linearity of the flow motion in the chamber and in the exhaust gas pipe. It was, however, possible to derive a scaling law for the damping in function of the gas temperature. A scaling law for the dependency of the damping on the length of the exhaust gas pipe could be also derived but its prediction is less accurate [14].

There is a possibility to determine the integral value of the damping ratio by one measurement e.g. at the resonance frequency predicted by the undamped Helmholtz resonator model. This is, however, feasible only if the combustion system already exists. In order to determine the value of the damping factor in the design stage numerical simulation should be carried out.

The present thesis describes the numerical investigation of a combustion chamber, as a single Helmholtz resonator, and a configuration of a burner plenum and a combustion chamber, as two coupled Helmholtz resonators. The reliability of the numerical method was tested by validating the results with the experimental data. Furthermore by means of the computations the main mechanism responsible for the damping could be identified.

# 3 Methodology

In recent years the rapid increase of computer power has made Computational Fluid Dynamics (CFD) a more and more widely used development and research tool. CFD is very useful for flow situations which are not accessible for measuring, for toxic flows etc. In the industry a faster turn around in the development can be realized. It makes possible a cheaper prototype testing. Furthermore measurements give information only in one point or in one plane from limited variables at the same time. If a plane is measured (e.g. with PIV) a huge amount of data is generated for post processing. CFD gives solution in each point of the domain of the simulation, all variables are available simultaneously. The development of the solution in time can be tracked easily. Despite of these advantages the completion of a CFD simulation with usable results is a great challenge. Depending on the hardware resources approximations and simplifications must be made for instance in the geometry or in the computation of turbulence. The choice of the numerical scheme, the choice and setup of the boundary conditions, the generation of the computational grid etc. provide a lot of opportunity to make mistakes. It is very important to gain experience by computing different test cases as much as possible.

At the Department of Fluid Machinery of the University of Karlsruhe an extensive experience with the development and use of very complex software packages for numerical flow simulation is available [28], [77], [78], [81]. The simulations of this work are carried out with SPARC (Structured Parallel Research Code) [78]. SPARC is an in-house developed block-structured Finite Volume research code for the simulation of compressible subsonic, transonic or supersonic flows [21], [22], [79], [83], aero-acoustic [89], [102] and transitional flows [86], [88]. Further research fields that can be touched with this code are simulations of fluid-structure interaction [101], multiphase flows and fully compressible combustion [84], [90]. SPARC is parallelised with the Message Passing Interface (MPI) [34] based on the domain decomposition method. It is very efficient on workstation clusters and supercomputers. Apart from the classical Reynolds-averaged Navier-Stokes Simulation (RANS) with a number of turbulence models like Baldwin/Lomax model, Spalart/Allmaras one equation model and a few two-equation $k$-$\tau$ models, it has been extended to Large Eddy Simulation (LES) since more than ten years. The Direct Numerical Simulation (DNS) as a special case of LES has been used for fundamental research for example in the PhD thesis of Franke [26].

Methodology

## 3.1 Governing Equations

The compressibility of the gas is very important in the pulsating combustion system despite of the low Mach number. The gas volume in the chamber acts as a spring as it was mentioned in Section 2. It will be shown in the results that in the range of the resonant frequency there is a backflow through the exhaust gas pipe. This means inflow into the chamber through both the inlet nozzle and the exhaust gas pipe at the same time.

The viscosity has a major role in the damping. The results will show that the oscillating viscous boundary layer is the main source of damping.

In order to capture all the physics in the combustion systems the fully compressible Navier-Stokes equations have to be solved. These can be written in integral form for the mass, momentum and energy conservation:

$$\int_V \frac{\partial \rho}{\partial t} dV + \int_S \rho \vec{v} \cdot \vec{n} dS = 0 \tag{3.1}$$

$$\int_V \frac{\partial (\rho \vec{v})}{\partial t} dV + \int_S \rho \vec{v}(\vec{v} \cdot \vec{n}) dS = -\int_S p\vec{n} dS + \int_S \vec{n} \cdot \mathbf{T} dS + \int_V \rho \vec{f} dV \tag{3.2}$$

$$\int_V \frac{\partial (\rho E)}{\partial t} dV + \int_S \rho E(\vec{v} \cdot \vec{n}) dS = -\int_S p(\vec{v} \cdot \vec{n}) dS + \int_S \vec{v}(\vec{n} \cdot \mathbf{T}) dS + \int_V \rho(\vec{v} \cdot \vec{f}) dV$$
$$- \int_S \vec{n} \cdot \vec{q} dS \tag{3.3}$$

Here $E$ represents the total specific energy, hence the sum of the inner and kinetic energy:

$$E = e + \frac{1}{2}\vec{v}^2. \tag{3.4}$$

In order to close the system of equations the thermodynamic and caloric equations for perfect gas is used. The equation of state then reads:

$$p = \rho R T \tag{3.5}$$

and the internal energy is only function of the temperature:

$$e = c_V T. \tag{3.6}$$

In Equation (3.3) $\vec{q}$ is being the energy flux vector, which is assumed to express only the molecular energy transport. This can be described by Fourier's law:

$$\vec{q} = -k\nabla T \tag{3.7}$$

with $k$ the thermal conductivity and $T$ the absolute temperature.

Assuming that the fluid is Newtonian the viscous shear stress tensor in Equations (3.2) and (3.3) reads:

$$\mathbf{T} = 2\mu\mathbf{D} - \frac{2}{3}\mu(\nabla\vec{v})\cdot\vec{I} \tag{3.8}$$

where the strain rate tensor $\mathbf{D}$ is defined as:

$$\mathbf{D} = \frac{1}{2}\left[(\nabla\vec{v}) + (\nabla\vec{v})^\mathrm{T}\right]. \tag{3.9}$$

In Equation (3.8) $\mu$ is the dynamic viscosity of the fluid. It is strongly influenced by temperature, which can be described by the widely used Sutherland's formula:

$$\mu = \frac{1.46 \cdot T^{3/2}}{T + 110.4} \cdot 10^{-6}. \tag{3.10}$$

On the right hand side in Equation (3.2) and Equation (3.3) $\vec{f}$ is a body force which can be e.g. the gravitation, or the modelled force of roughness or an extra force which should drive the flow.

## 3.2 Turbulence Modelling

The equations above describe generally fluid motions. If the domain of interest is discretized that all gradients are resolved adequately, the flow motions are simulated accurately, depending only on the numerical scheme. Compared to laminar flows turbulent motions produce higher gradients of the variables. If these gradients are still resolved the simulation is called Direct Numerical Simulation (DNS). As the computational effort is rising with $Re^3$ DNS is restricted today to academic research on supercomputers and to Reynolds numbers much lower than the range of engineering interest. In order to compute highly turbulent flows with lower computational effort, the modelling of turbulent motions, at least in part, is needed.

The great idea of Reynolds [110] splitting the flow variables in mean and fluctuating part provided a useful technique for modelling all the turbulent motions and founded a group of models called Reynolds-averaged Navier-Stokes (RANS). For detailed description of the method it is referred to Wilcox [146]. It will be briefly illustrated here that the splitting leads to appear of new variables, the Reynolds stresses, and therefore to a closure problem. One possibility is to close the system by means of the Boussinesq eddy-viscosity approximation [11]. Boussinesq assumed that turbulent stresses act as viscosity stresses, which implies that the turbulent stresses are proportional to the mean strain stresses. The turbulent viscosity can be calculated in the simplest way by an algebraic expression e.g. as by the model of Baldwin and Lomax [7]. This model is suitable for high-speed external flows with thin attached boundary layer. Advantages are its robustness and the fast computation. Complex flows with separation should be computed with more advanced models, which solve one or two transport equations of turbulence variables to compute the eddy viscosity. The most popular one-equation model proposed by Spalart and Allmaras [129] solves a transport equation for a viscosity-like variable. Two-equation models solve a transport equation for the turbulent kinetic energy $k$ and for a further variable representative for turbulence. This can be the dissipation of the turbulence $\varepsilon$, the specific dissipation rate $\omega$ or the turbulence dissipation time $\tau$. About advantages and disadvantages of the different models it is referred again to Wilcox [146]. It is very difficult to decide, which model should be used. There are many comparative works in the literature; they investigate different models for a given test case. But the pros and cons of a model can change slightly from test case to test case. It is more important to see, if the complexity of the flow situation increases the failure of the models increases also. If the time-dependency of the mean flow is increasing the unsteady RANS (URANS) simulations provide better results, however, at the cost of rapid increasing of the computational effort. If the turbulence in the computed case is not restricted to a thin boundary layer, the assumption of isotropic turbulence, used for eddy-viscosity models, induces further failure. In turbulent flows the small structures are more isotropic whereas the large structures are more anisotropic. Based on this the Large Eddy Simulation (LES) is the approach, which resolves the large scales and models only the isotropic small scales. Hence LES is a compromise between DNS and URANS. As the computational effort of LES is considerable, the (U)RANS method still dominates in industrial applications. RANS can provide results in hours or days on workstations or on workstation clusters for relative large computational domain and high Reynolds numbers. In order to overcome its shortcomings it will be extended to complex flows with additional modelling (e.g. the curvature of the mean flow [131]). In order to use the advantages of (U)RANS it can be combined with LES as in the Detached Eddy Simulation (DES) [130], in hybrid

RANS/LES [29], in Very Large Eddy Simulation (VLES) [24], [132] or in the Partially-Averaged Navier-Stokes model (PANS) [33].

**Large Eddy Simulation**

The turbulent energy in the flow is distributed among different scales of the turbulent motions. It is transferred mainly from the large structures to the smaller. The large scales are very important as they are generally much more energetic. Their size and strength make them by far the most effective transporters of the conserved properties. The large scales are approximately inviscid, coherent and anisotropic. The size of the smallest scales, the so-called Kolmogorov microscales, is limited by the viscosity. As the viscosity is dominating in this range these structures are dissipative and isotropic. By the prediction of flows in complex geometries, where large, anisotropic vortex structures dominate, the statistical turbulence models often fail. The Large Eddy Simulation approach is for such flows more reliable and more attractive as it allows more insight into the vortex dynamics. In recent years the rapid increase of computer power has made LES accessible to a broader scientific community. This is reflected in an abundance of papers on the method and its applications.

LES is an approach to simulate turbulent flows based on resolving the unsteady large-scale motion of the fluid while the impact of the small-scale turbulence on the large scales is accounted for by a sub-grid scale model. Hence the modelling error is reduced significantly.

The formulation of the equations for LES is quite similar to them of RANS. The difference is that whereas the mean values are built for RANS through averaging in time, for LES the averaging is made in space and they are called filtered values. Further difference between LES and RANS is the effect of the grid refinement. If the mesh is refined the solution of RANS converges to a mesh independent solution. Such a study is easily available with the full multigrid method, which is implemented in SPARC. It is important to mention that if the grid is too fine, the simulated flow can be unstable i.e. unsteady. If the flow is unstable we get no convergence any more. It is great in LES that the modelling error is decreasing with smaller grid size. The grid refinement of LES yields to DNS. Further refinement is useless.

The small, unresolved scales are smaller than the cell size therefore they are called subgrid-scales (SGS). As these scales are quite isotropic the modelling error is much smaller than for RANS. Similar to RANS the closure problem can be solved by different approaches. The use of the Boussinesq eddy-viscosity approximation is here also a good solution. The earliest and most commonly used subgrid-scale model is the algebraic model of Smagorinsky [128]. It is based on the eddy viscosity concept and it computes the eddy viscosity from the strain rate of the resolved field:

$$v_t = (2 \cdot C_S \cdot \Delta)^2 |\overline{S}|. \qquad (3.11)$$

It has been found that the Smagorinsky-constant $C_S$ is slightly varying depending on the Reynolds number or on other non-dimensional numbers and on the computed flow situation. The filter length scale $\Delta$ can be computed differently [29]. In SPARC this relation is used:

$$\Delta = \sqrt[3]{\Delta x \cdot \Delta y \cdot \Delta z}. \qquad (3.12)$$

Turbulence is a highly complex phenomenon, therefore it is not a big surprise that the simple Smagorinsky model fails to cover all types of turbulent flows. However it builds the base for further models (e.g. dynamical Smagorinsky model [30], HPF Smagorinsky model [133]) and for further development, which is today a dynamically expanding research area. For a more complete description of LES, see the book e.g. of Geurts [32] or Sagaut [117].

The modelling of the effect of the non-resolved scales by the calculation of an eddy-viscosity is called explicit method. A further possibility to model the effect of the subgrid-scale turbulence is to use the numerical scheme of artificial dissipation. The modelling of the effect of the subgrid-scale turbulence happens here implicitly.

In SPARC the central differencing scheme (CDS) is applied for the spatial discretization. Due to the mathematical formulation of CDS, odd-even decoupling can occur in the equation system [41]. High gradients, typically present in shock waves, at phase boundaries and stagnation points, cannot be captured cleanly with CDS. Upwind differencing schemes (UDS), on the other hand, can handle discontinuities without oscillations, as they are inherent dissipative. In order to overcome this problematic in CDS formulation, Von Neumann and Richtmyer [145] introduced the concept of artificial dissipation or artificial viscosity. Artificial dissipation schemes are mainly used in SPARC for RANS/URANS simulations.

The application of the artificial dissipation as an implicit subgrid-scale model for Large Eddy Simulation was proposed firstly by Boris et al. [10] and was called as the monotone integrated Large Eddy Simulation (MILES) approach. This method is used in SPARC with the Matrix artificial dissipation scheme (Swanson and Turkel [134]) based on the non-linear SLIP method (Tatsumi et al. [135]).

## 3.3 Full Multigrid Method

The multigrid strategy is a very efficient convergence acceleration technique. It was developed for the solution of elliptic problems [13] and later was extended to hyperbolic problems [53], [98]. The basic idea is to introduce a sequence of coarser grids and use them to speed up the propagation of the fine grid corrections, resulting in a faster expulsion of disturbances. The coarser auxiliary meshes are obtained by doubling the mesh spacing, i.e. omitting every second gridline. This can be easily achieved on structured meshes. In order to provide a well conditioned starting solution for the fine mesh the Full Multigrid (FMG) method is used. The FMG is analogous to grid sequencing, except that multigrid cycles are performed on each coarser grid. In SPARC a V-type cycle with subiterations is used as a multigrid strategy [78]. Further convergence acceleration techniques are implemented in SPARC as local time stepping and implicit residual smoothing [78].

## 3.4 Dual Time Stepping

Explicit time integration methods are conditionally stable. The stability is expressed by the Courant-Friedrichs-Lewy condition (CFL), which requires that the domain of dependence of the numerical scheme must at least contain the region of dependence of the original differential equation i.e. CFL$\leq$1 [146]. Hence, the time step for the explicit scheme is determined by the smallest cell with the highest velocity. In compressible computations for wall-bounded, turbulent, subsonic flows with low Mach number this is mostly the first cell at the wall if $y^+\leq1$ condition is applied (see Section 3.6.1). Here the velocity is approximately the speed of sound. The propagation of some information with the speed of sound in the viscous sublayer is not exactly the focus of our interest.

Implicit methods, although expensive in computation, have less severe stability bounds (classical stability analysis shows unconditional stability but in practice on nonlinear problems bounds are encountered). The extra work required for an implicit scheme can be reduced significantly by the convergence acceleration techniques. By means of an implicit method we can decide which information should be resolved. Depending on the stretching of the mesh normal to the wall the time step can be 15-20 times higher than for the explicit scheme without any impact on the spatially resolvable turbulent scales. Implicit method makes the choice of higher time step possible. But care must be taken, if the time step is too high, turbulence scales, which could be resolved spatially, will be temporally unresolved.

In SPARC the implicit Dual Time Stepping method is implemented. In the dual-time procedure the physical time layer is used to track the physical variety of the flow, while a pseudo-time layer is used to reach iteratively every physical time step [154].

## 3.5 Mesh generation

Mesh is the set of support points for the equation system arranged in space. Some people treat the mesh generation as an imperative, annoying duty and underestimate the importance of the mesh quality. However, the quality of the mesh has a large influence on the solution.

In SPARC body-fitted block-structured mesh is used. Computation on block-structured meshes can be much faster than on unstructured meshes. Parallelized by domain decomposition, the blocks are distributed among the processors. Block-structured mesh can be better used for complex geometries than structured meshes. If the complexity of the geometry is very high the saving of control volumes (CV) is very hard and the use of an unstructured mesh becomes more favourable.

In order to get reliable results some simple main rules of thumb are constructed for the mesh generation. The discretization error can be kept within acceptable limits if these are kept. The angle between the adjacent sides of neighbouring cells ($\alpha$ in Figure 3.1) should be smaller than 30°. The aspect ratio for cells ($L_1/h$ in Figure 3.1) is ideally unity for LES, but this can lead to a huge number of CVs. An optimum must be found. The aspect ratio depends also on the form of the resolved structures.

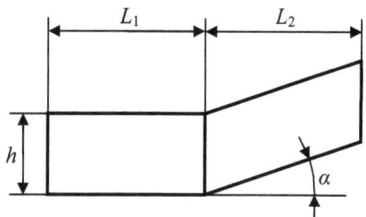

**Figure 3.1:** Mesh generation rules

The stretching of cells ($L_2/L_1$ in Figure 3.1) is limited to some percents in the crucial regions but it should be not too high in the unimportant regions because it can cause reflections. The mesh should be fine in regions of steep gradients and coarse in regions of low gradients. Latter is important in order to save computational time. The meshes for the investigations were generated in ICEM CFD, the commercial software from ANSYS [147]. In SPARC a parallelised smoothing algorithm and quality check is available to help improve the quality of the mesh.

## 3.6 Boundary Conditions

The numerical simulations are always performed for a well defined flow problem. At the interface between the domain of interest, where the simulation is carried out, and the surrounding area boundary conditions for the differential equations must be described. The boundary conditions must be placed where the flow properties are well known as the boundary conditions determine the solution. The boundary conditions used for the simulations and implemented during the investigations of the combustion systems are described in this section. For more details about boundary conditions it is referred to [41].

### 3.6.1 Wall

The most widely used condition is the wall. The simplest is a no-slip, adiabatic wall which is imposed by setting the flow velocity and the gradient of the temperature to zero at the solid surfaces. The pressure is calculated from the normal momentum equation. The shape of the wall is also a very important property of this boundary condition. The geometry of a body is obviously well known but in the most cases it is approximated if its complexity is too high. The degree of approximation is influenced by the computation resources but it must be handled with care which parts of the complex details are omitted.

**Near-Wall Treatment of Turbulence**

Modelling turbulence near to walls is an important issue for the simulation. The statistical models can be split in two groups. The so-called low-Re models resolves the entire flow normal to the wall. As the steepest velocity gradient is at the wall, a very high eddy viscosity should be predicted here. In order to avoid this unphysical issue Van Driest damping function is used in the vicinity of the wall. When using

## Methodology

low-Re models, the spatial resolution at the wall will be very fine. In general, for the first cell at the wall the size expressed in wall units should be $y^+ \leq 1$. This leads to a very large number of points, which increases the computational effort. These are, however, necessary for flows with separation and reattachment of the boundary layer. For LES this method is used basically. At higher Reynolds numbers the low-Re method yields, however, to very small cells. For RANS models this generates very narrow and very long cells, indicating a very poor ratio of length to height/width (aspect ratio). This is reflected then in the form of large eigenvalues of the equation system and in degradation of the convergence. In LES the allowed aspect ratio is much smaller than for RANS simulations. This leads to a very large number of computational cells in the entire computing field. A further drawback exists for explicit time integration, since the time step is determined by the smallest cell.

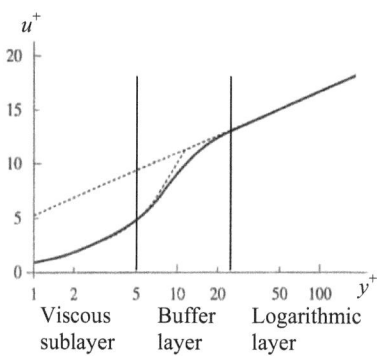

**Figure 3.2:** Distribution of the mean velocity of the turbulent boundary layer

Contrary in the high-Re models neither the viscous sublayer nor the buffer layer is computed (see Figure 3.2). These models use a wall function to bridge these small regions of high velocity gradient. The base of wall modelling is the universal logarithmic law of the wall:

$$u^+ = \frac{1}{\kappa} \ln y^+ + C \qquad (3.13)$$

with

$$u^+ = \frac{U_t(y)}{u_\tau}, \quad y^+ = \frac{u_\tau \cdot y}{\nu} \quad \text{and} \quad u_\tau = \sqrt{\frac{|\tau_w|}{\rho}}, \qquad (3.14)\,(3.15)\,(3.16)$$

20

where $y$ is the normal distance to the wall, $u_\tau$ is the shear stress velocity, $U_t(y)$ is the velocity tangential to the wall and $\tau_w$ is the shear stress at the wall. The value of the so-called Karman-constant $\kappa$ is approximately 0.41. The constant $C$ is approximately 5.0 for smooth walls and changing with roughness or with pressure gradient in flow direction.

If the Equation (3.13) is used, the first computational point should be at $y^+ \approx 30$. For complex flows this condition is hard to keep therefore the description of the mean profile of the turbulent boundary layer was extended to all regions. Further extension was needed to be able to deal with separation and reattachment [127], [143]. In order to save on computational cells for LES the adaptation of the wall model is today a dynamic research field [95], [96].

Which approach can be used for pulsating flows is discussed in Section 4.2.

**Wall Roughness – The Discrete Element Method**

It has been well known for many decades that the roughness of a wall has a big impact on the wall stresses generated by a fluid flow over that surface. It is of great importance in many technical applications to take this property into account in the simulations, for example for the accurate prediction of the flow around gas turbine blades. The modelling of wall roughness has traditionally been done using the log-law of the wall [1], [3], [150]. There the wall-law is simply modified to account for the roughness effect. Unfortunately the log-law of the wall is not general enough to be applied for complex flows, especially flows with separations and strong accelerations or decelerations (as the flow in the resonator neck).

Another modelling has been proposed and used in the past which is based on a modification of the turbulence model close to the wall [5], [18], [92]. Here for example the non-dimensional wall distance is modified to account for the wall roughness in the algebraic Baldwin-Lomax model or the Spalart–Allmaras one–equation model. These are, however, restricted for Reynolds Averaged Navier-Stokes (RANS) calculations. A more general modelling is the Discrete Element Method (DEM) first proposed by Taylor *et al.* [136] which can be used for all types of simulation tools. Since our main focus is on Large Eddy Simulation this approach seems to be most appropriate. The main idea of the method is to model the wall roughness effect by including an additional force term into the Navier-Stokes equations by assuming a force proportional to the height of virtual wall roughness elements.

# Methodology

The DEM models the effect of the roughness by virtually replacing the inhomogeneous roughness of a surface by equally distributed simple geometric elements for example cones (Figure 3.3).

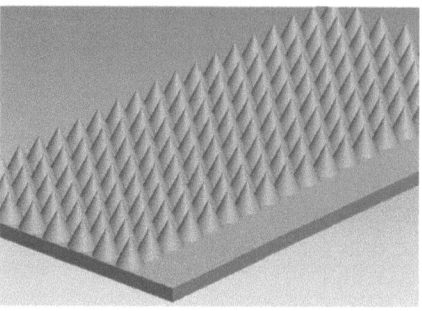

**Figure 3.3:** Distribution of the simulated cones

The height and the number of the cones per unit area are the parameters of the model. Since the force of a single cone on the fluid can be approximated by known relations it can be used to simulate the drag of the roughness onto the flow. While Taylor *et al.* have modelled the influence of the discrete element by assuming blockage effects onto the two-dimensional Navier-Stokes equations, Miyake *et al.* [94] have generalized this idea by explicitly accounting for the wall drag force $f_i$ as source terms in the momentum and energy equations:

$$\frac{\partial \rho}{\partial t} + \frac{\partial \rho u_i}{\partial x_i} = 0 \tag{3.17}$$

$$\frac{\partial \rho u_i}{\partial t} + \frac{\partial \rho u_i u_j}{\partial x_j} = -\frac{\partial p}{\partial x_i} + \frac{\partial}{\partial x_j}\left[\mu\left(\frac{\partial u_i}{\partial x_j} + \frac{\partial u_j}{\partial x_i} - \frac{2}{3}\frac{\partial u_k}{\partial x_k}\delta_{ij}\right)\right] + f_i \tag{3.18}$$

$$\frac{\partial \rho E}{\partial t} + \frac{\partial \rho u_i E}{\partial x_i} = \frac{\partial (q_i - p u_i)}{\partial x_i} + \mu\left[\frac{\partial u_i}{\partial x_j}\left(\frac{\partial u_i}{\partial x_j} + \frac{\partial u_j}{\partial x_i}\right) - \frac{2}{3}\left(\frac{\partial u_i}{\partial x_i}\right)^2\right] + f_i u_i \tag{3.19}$$

Miyake *et al.* assume the specific drag force $f_i$ to be proportional to:

$$f_i = c_D \cdot \frac{1}{2}\rho u_i^2 \cdot \frac{A_C}{V} \tag{3.20}$$

Here $A_C$ is the projected surface of the cone in the flow direction, $u_i$ is the velocity component in the appropriate direction, $c_D$ the drag coefficient of the cone and $V$ is the volume of the cell. In the experiments $c_D$ varies between 0.2 and 1.5. For the present computations $c_D$=0.5 was chosen.

The effect of the roughness onto the wall shear stress must also explicitly be accounted for by:

$$\tau_{wi} = \mu \frac{\partial u_i}{\partial n} + \frac{f_i \cdot V}{A} \qquad (3.21)$$

Here $A$ is the projected surface of the cone onto the wall and $n$ is the normal distance from the wall.

The implemented method has been validated and calibrated by Bühler [17] at the flat plate test case using the experimental findings of Aupoix and Spalart [5] (Figure 3.4). Here the height of the virtual cones $h_r$ is compared with sand grain roughness $k_s$.

The prediction of the roughness influence on a high pressure turbine blade has been investigated next and compared with experiments of Hummel and Lötzerich [47]. It was found that the method works very satisfactorily when used with the Spalart et al. one-equation turbulence model for RANS.

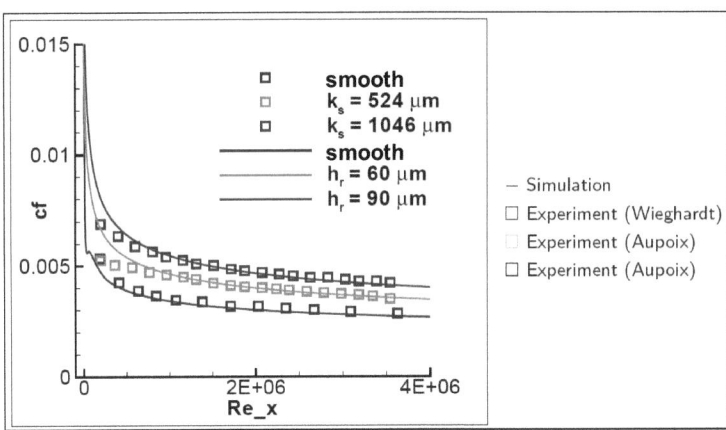

**Figure 3.4:** Flow over the flat plate. Drag coefficient versus $Re_x$

## 3.6.2 Non-Reflecting Boundary Condition

Boundary conditions are used to prescribe the solution of the equations on the boundaries of the computational domain. Dirichlet type boundary conditions describe

## Methodology

the value of a variable, while von Neumann type boundary conditions describe the value of the gradient of the variable. There are also mixed type boundary conditions.

Dirichlet type boundary conditions produce a partial reflection of disturbances, which are produced in the flow domain and can reach the boundary. Simplest example is the pressure fluctuation in the vortices of the wake behind a turbine blade. If these vortices reach undamped the pressure outlet boundary condition pressure fluctuations will be reflected back to the flow domain causing unphysical phenomena. The damping is available physically by viscosity if the outlet is enough far away. The damping can be increased artificially by stretching of the mesh and e.g. by adding of an extra artificial viscosity in a layer at the boundary. In far-field weekly non-reflecting boundary condition can be constructed by means of the Riemann variables [41]. Sponge layer type non-reflecting boundary conditions are more popular [23], [27], [43].

The series of numerical investigations showed that under the numerous possible error sources which affect the solution (e.g. commutation error, coarse grid, large aspect ratio of cells) the unphysical reflection of the pressure signal on the boundaries of the computational domain is significant. In the case of solving the compressible equations for an internal flow these reflections must be damped by an effective method.

In SPARC the non-reflecting condition of Freund [27] was implemented. He proposed the method for external flows to damp the reflections of acoustic waves efficiently. The method is based on a buffer layer technique, which use additional convective and dissipative terms in a layer at the inlet/outlet boundary. The generic simplified one-dimensional equation reads:

$$\frac{\partial \Phi}{\partial t} + A(x)\frac{\partial \Phi}{\partial x} + \frac{\partial (u\Phi)}{\partial x} = -\sigma(x)\left[\Phi - \Phi_{target}\right] \qquad (3.22)$$

Freund proposed an additional convective term $A(x)$ with a maximum value of $A_0=1.15 \cdot c$ which accelerates the subsonic flow above the speed of sound changing locally the characteristic of the equations from elliptic to hyperbolic type. In the inlet buffer layer it causes stopping of the disturbances travelling upstream, while in the outlet buffer layer it advects out any disturbances travelling downstream. More details of the method are described in [27].

The first implementation of this method has shown no significant improvement of the solution by calculating the transient transonic flow around the VKI turbine blade

# Methodology

(Magagnato *et al.* [87]). Therefore some modification of the method was applied to make it applicable for internal flows also.

For compressible subsonic flow four boundary values must be given at the inlet and one must be extrapolated from the flow area. At the outlet one must give one boundary value and extrapolate four others. The prescribed information at the outlet has to propagate from the boundary upstream the flow in order to get the correct solution in the whole domain. If the type of the equations is changed from elliptic to hyperbolic in the buffer layer at the outlet, the prescribed information will be lost for the entire domain.

In order to solve this problem, the method of Freund was modified. The flow will be accelerated in the buffer layer with the additional convection term so that the flow remains subsonic and the character of the equations remains elliptic in the whole domain ($A_0=0.95 \cdot c$). This permits a slight reflection of pressure disturbance in the buffer layer at the outlet boundary, but they travel very slowly upstream and the additional damping term dissipates the fluctuations. Likewise acts the additional damping term in the buffer layer at the inlet boundary. Here the disturbances travelling upstream will be strongly decelerated and effectively damped. The turbulent fluctuations imposed at the inlet boundary travel downstream through the layer in short time because of the acceleration, and reach the physical domain almost unaffected by the damping. At the boundary of physical domain the flow speed achieves its proper value. In both buffer layers the acceleration/deceleration is not abrupt but smooth.

The one dimensional test case described in Rachwalski *et al.* [108] revealed 97% improvement compared to reflecting boundary conditions i.e. the reflection has been reduced from about 60% down to less than 2%. With the new modifications this maximal damping cannot be reached but the reflection is still reduced to about 5%.

### 3.6.3 Mass Flow Rate Inlet

In the investigations the flow in the combustion chamber is non-reacting and excited with a sinusoidal mass flow rate. The latter property necessitates the use of a relative complex boundary condition at the inlet for the LES.

The first implementation of mass flow rate inlet boundary condition was based on the total pressure boundary condition. The pressure was varied to obtain the desired mass flow rate. This indirect connection allows the development of a boundary layer at the inlet. This approach is applicable to steady flow simulations but its dynamics (because of the compressibility of the fluid) makes it inapplicable for unsteady flows, especially with pulsating inlet.

Therefore the approach of the velocity profile boundary condition was adopted for the mass flow rate inlet. Here the components of the velocity vector and the temperature is prescribed by the boundary condition, the density is extrapolated from the computational domain. The mass flow rate is measured at the inlet cross section in each time step and the velocity is scaled based on the fraction of the desired and the obtained mass flow rate. The distribution of the velocity is very important close to the wall, i.e. in the boundary layer. This can be either prescribed by the user if known from e.g. experiment or extrapolated from the computational domain.

The extrapolation of the distribution of the velocity is applicable if the flow near to the inlet boundary is not disturbed (e.g. there is no diversion near to the inlet boundary) or the rate of the pulsation is moderate. This method leads to a fully developed channel or pipe flow.

For higher rate of the pulsation or with prescribed boundary layer thickness the use of a prescribed distribution of the velocity is indispensable.

### *3.6.4 Precursor Type Inlet Condition*

It is known from experiment that the inflow into the combustion chamber and into the burner, respectively, is highly turbulent. It is very important to investigate the effect of the turbulence intensity of the inflow on the resonant characteristics of the combustion system. Therefore, an appropriate inlet condition has to be employed. Since a pulsating mass flow rate has to be produced for the investigation of the resonant characteristics of the burning system, the use of a complex inlet condition is required.

It is widely accepted that the specification of realistic inlet boundary conditions plays a major role in the accuracy of a numerical simulation [59], [62], [76].

In the case of a Reynolds averaged Navier-Stokes (RANS) simulation it is relatively simple to define proper data at the inlet. The profile of the mean velocity and the turbulence variables have to be prescribed simply.

For LES and DNS a time dependent series of data for all the velocity components and in the case of a subsonic compressible flow for a further thermodynamic variable (e.g. temperature) are required. In order to produce time dependent turbulent inflow for LES and DNS there are some methods in the literature.

The simplest procedure for specifying turbulent inflow conditions is to superpose random fluctuations on a desired mean velocity profile. The shortcoming of this boundary condition is that the turbulence spectrum of the incoming flow is not recovered. Since these fluctuations are not coherent and have no correlations in space and time, they do not reflect real turbulent fluctuations. Turbulent coherent large-scale structures initiate the cascade of turbulent kinetic energy from large to small scales. The inflow fluctuations are inserted with the frequency of the time step hence they do not contain the whole energy spectrum of turbulence. As a result, the fluctuations have more energy in the high wave numbers and dissipate very fast without sustaining or initiating real turbulence.

The most accurate method that retains a degree of generality consists of obtaining inflow data from a precursor simulation [23], [59], [76], [118]. Here an auxiliary simulation of wall-bounded flow produces fully developed turbulent flow data with periodic conditions in the streamwise direction. Spatially developing turbulent boundary layer data by means of a rescaling method is also possible [23], [76], [118].

In the precursor region the desired mass flow rate can be obtained with a prescribed pressure gradient or with a forcing term. The use of a body force has two advantages. Unlike the use of a pressure gradient, a better homogeneity of the flow can be obtained, and the adjustment of the mass flow rate is fairly faster.

A body force is added to the streamwise momentum and energy equations in order to compensate the wall-friction forces and to obtain the required mass flow rate.

Lenormand *et al.* [66] described a method for the calculation of the body force for a compressible channel flow. The method is constructed to maintain a constant mass flow rate which should be previously obtained already by the initialization. Furthermore the initialized velocity profile is of great importance since the wall shear stress is used to predict the driving term.

In the general case it is very laborious to construct proper initialization values for each computation therefore a generally working method was needed. On the other hand in SPARC a full multigrid method is implemented, which is very useful in the case of complex geometries, where a proper initialization is almost impossible. If the precursor simulation is coupled to the main domain, the flow in the precursor region develops simultaneously with the main domain from the initialization in each multigrid level. Additionally, in the case of the combustion chamber the method described in [66] is difficult to realise because of the required pulsation of the mass flow rate. In order to solve these problems a controller based on the instantaneous mass flow rate difference is constructed. The controller has to handle the flow in the precursor region as a dynamical system consisting of a mass, spring (compressibility) and weak, nonlinear damping with hard definable parameters. In order to prevent overshootings of the mass flow rate the forcing term is maximized with $g_{max}$. The streamwise component of the forcing term at the time step $n+1$ is read:

$$g_1^{n+1} = g_1^n + g_{max} \cdot MIN\left(1, \left|\frac{\Delta \dot{m}}{\dot{m}_{ref}}\right|\right) \cdot SIGN(\Delta \dot{m}) - a^n \qquad (3.23)$$

with

$$\Delta \dot{m} = \dot{m}^n - \dot{m}^{n-1}, \quad a^n = \frac{1}{\langle \rho \rangle 2hL_y} \frac{\Delta \dot{m}}{\Delta t}. \qquad (3.24), (3.25)$$

The maximum value of the forcing term can be approximated by the amplitude and the frequency of the excitation mass flow rate:

$$g_{max} \approx 2\pi \cdot \frac{1}{\langle \rho \rangle 2hL_y} \hat{\dot{m}} \cdot f_{ex} \qquad (3.26)$$

In the case of the initialization, the exaggerated oscillations can be avoided by means of an exponential type driving function, and the prescribed mass flow rate can be reached much faster. This is demonstrated with a steady test case of laminar flow, where a constant mass flow rate is to be realised (Figure 3.5 and 3.6).

In the compressible channel flow the shear stresses produce heat which results in decreasing of the density along the channel. In terms of the mass conservation with decreasing density the velocity has to increase. In this manner, unlike an incompressible channel flow, the state of the fluid has to be described more accurately if periodicity is used in streamwise direction. For the simulation the total temperature is held constant in the channel centreline at the inlet boundary. This can be applied, if the total temperature fluctuation is negligible as discussed in [12].

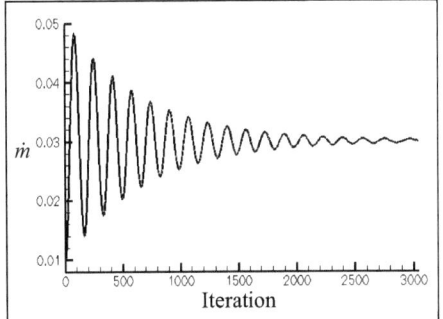 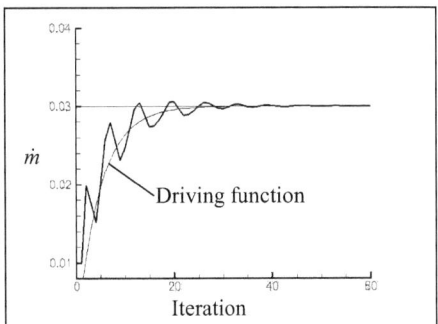

**Figure 3.5:** Control of the mass flow rate based on the instantaneous mass flow rate difference

**Figure 3.6:** Control of the mass flow rate completed with a driving function

As validation test case the turbulent channel flow of Reynolds number $Re_\tau=395$ is simulated. For the discretization in space a second order accurate central difference scheme was used. In time a second order dual time stepping scheme was applied. The time step was set to $\Delta t=10^{-5}$ s. A full multigrid method is used to accelerate the convergence in the inner iterations. The use of the full multigrid method allows also the investigation of the effect of the grid refinement.

The computations were performed using the Smagorinsky-Lilly subgrid scale model. The Smagorinsky coefficient was set to $C_S=0.13$. A computation with $C_S=0.1$ gave approximately the same results for the turbulent channel flow.

The target Reynolds number is $Re_\tau=395$, the Mach number based on the bulk velocity and sound speed at the walls is $Ma=0.3$. The mass flow rate is held at a constant value.

The computational domain of the precursor simulation is $(2\pi h, 2h, \pi h)$ in $(x, y, z)$, where $x$ is the streamwise direction, $y$ the wall normal direction, and $z$ the spanwise direction. The grid resolution is $(64, 160, 64)$ in $(x, y, z)$. The grid size is in the streamwise direction $\Delta x^+=25$, in the spanwise direction $\Delta z^+=12$ and in the wall normal direction $\Delta y^+=0.4$ at the wall and stretched up to $\Delta y^+=2$ at the centreline of the channel. A body force $g$ as discussed above is applied in this domain to obtain the prescribed mass flow rate (Figure 3.7).

Methodology

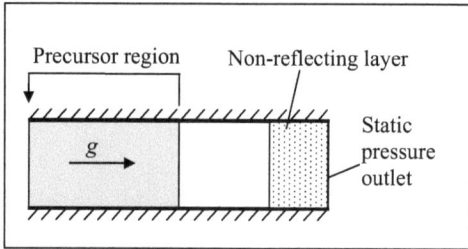

**Figure 3.7:** The sketch of the computational domain

Periodic boundary conditions are applied in the spanwise direction, while no-slip conditions are applied at the adiabatic walls.

Boundary conditions comprise always approximations furthermore a standard static pressure outlet is very stiff against oncoming waves like e.g. the turbulent vortices and produces reflections. Therefore an outlet condition at the end of the precursor domain is not applicable. Downstream of the precursor simulation a domain with the same geometry is coupled. This domain substitutes now also the main domain of interest. At the outlet boundary the static pressure condition is used. In order to avoid reflections the grid is stretched towards the outlet boundary and a non-reflecting layer is used as described in [86].

As soon as the statistical steady state was obtained the statistics were sampled till the mean value of the spanwise velocity component was less than 1% of the mean streamwise velocity. Additionally to time averaging, the statistics were averaged also over the $z$ direction (homogeneous direction).

The results of the simulation are compared to the DNS data by Iwamoto and Iwamoto et al. [51], [52] for the mean velocity and for the velocity fluctuations.

Figure 3.8 shows the mean profile of the streamwise velocity component. For the sake of the perspicuity only each fourth of the DNS data are plotted.

The LES results compare well with the DNS data although the LES overpredicts somewhat the mean velocity profile in the buffer layer and in the log-law region. Since the same result is obtained in the domain downstream of the precursor region, this seems not to be an effect of the precursor simulation. The overprediction can arise from the grid resolution, from the filtering or from the discretization scheme etc. It is not the aim of this work to investigate the source of this discrepancy.

Methodology

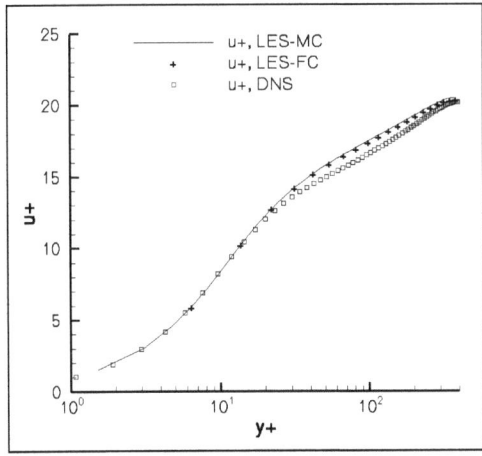

**Figure 3.8:** The mean velocity profile in wall units (only each fourth data are plotted in the case of DNS and LES-FC)

The difference between pressure gradient driven flow and body force driven flow is negligible small, as discussed by Huang *et al.* [44], and cannot produce such a deviation in the velocity profile.

In order to check the propriety of the precursor simulation, another investigation was more relevant. For the control of the forcing term the mass flow rate in each time step is needed. The instantaneous mass flow rate can be calculated in one cross section of the precursor region, e.g. at the inlet plane. In this case, if the flow becomes turbulent, the mass flow rate starts to oscillate, and therefore the forcing term oscillates strongly, too. For the reduction of this oscillation the mass flow rate in the precursor domain was averaged in the streamwise direction. This reduced the amplitudes but the oscillation remained. Therefore one computation was performed with a constant forcing term to investigate the effect of this oscillation.

Figure 3.8 shows that the results of the calculation with controlled force (LES-MC) do not deviate substantially from the calculation with constant forcing term (LES-FC). For the sake of the perspicuity only each fourth of the LES-FC data are plotted.

In the case of controlled forcing term (LES-MC) the maximal relative error of the mass flow rate is within 0.01%. The constant value for the forcing term was approximated with the integral of the oscillating forcing term on a certain time interval. Although the exact value was not caught the mass flow rate deviates from

Methodology

the desired $\dot{m}_{ref}=0.4\ kg/s$ within 0.5%. The temporal evolution of the averaged mass flow rate and the forcing term are plotted in Figure 3.9.

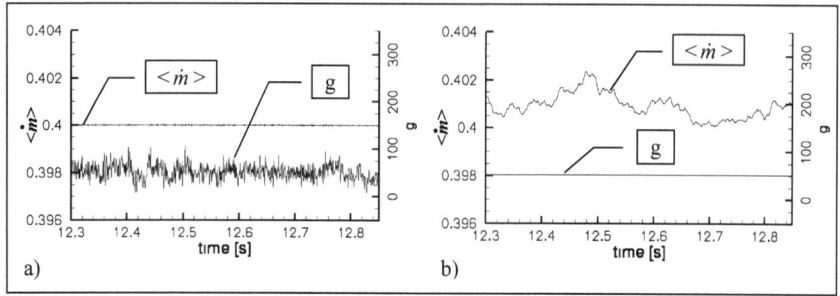

**Figure 3.9:** Temporal evolution of the averaged mass flow rate and forcing term: a) constant mass flow rate (LES-MC), b) constant force (LES-FC)

Figure 3.10 and 3.11 show the filtered RMS-fluctuations and the Reynolds shear stress. The LES results compare satisfactorily with the DNS data. A part of the deviation is obtained because the realized Reynolds number is slightly lower ($Re_\tau=380$) than in the DNS ($Re_\tau=395$). The results from LES-FC are not plotted in Figure 3.10 and 3.11 because of the negligible deviation from the LES-MC.

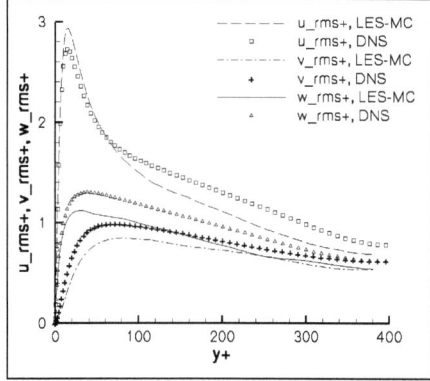 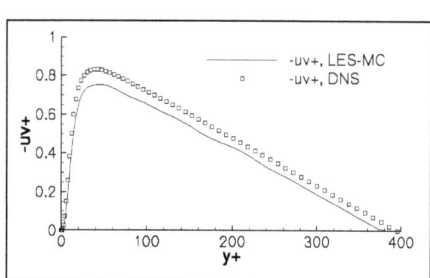

**Figure 3.10:** Filtered RMS velocity profiles in wall coordinates (only each fourth DNS data are plotted)

**Figure 3.11:** Filtered Reynolds shear stress profile (only each fourth DNS data are plotted)

This inlet condition with pulsation was tested on the single resonator. The results are described in Section 4.4.1.

### 3.6.5 Synthetic Eddy Method

Although the precursor simulation is the most accurate method to generate inflow turbulence, its drawbacks are the non-negligible additional computational cost, the restriction to wall-bounded flows and the highly complicated combination with a non-reflecting layer. The Synthetic Eddy Method was chosen to be implemented in SPARC, which do not need an additional computational region. It is applicable for all internal flows and the combination with non-reflecting condition is easier.

Recently Jarrin et al. [55] proposed an efficient technique that generates turbulent inflow data. The method performs well on any geometry and any kind of flow. The generated data exhibit very good physical properties. We extended this method for compressible flows and implemented it in SPARC. The method was tested on the spatially decaying isotropic turbulence.

The inflow generation method described in [55] is based on the classical view of turbulence as a superposition of eddies. Coherent structures will be generated over the inlet plane by means of spots. The spots are designed to produce correlation in space and time. A short description of the method is given here, more details can be found in [55].

In order to produce turbulence in an x-plane, the velocity field is constructed as:

$$u_j(y,z,t) = \bar{u}_j(y,z) + a_{ij} u'_j(y,z,t) \qquad (3.27)$$

with

$$a_{ij} = \begin{pmatrix} (R_{11})^{1/2} & 0 & 0 \\ R_{21}/a_{11} & (R_{22}-a_{21}^2)^{1/2} & 0 \\ R_{31}/a_{11} & (R_{32}-a_{21}a_{31})/a_{22} & (R_{33}-a_{31}^2-a_{32}^2)^{1/2} \end{pmatrix} \qquad (3.28)$$

where $R_{ij}$ is the correlation tensor which may be known *a priori* from experiments. The overbar denotes the mean component.

The provisional signal for the fluctuating velocity component $j$ reads:

$$u'_j(y,z,t) = \frac{1}{\sqrt{N}} \sum_{n=1}^{N} u^{(n)(j)}(y,z,t). \qquad (3.29)$$

The contribution of the spot $n$ is being:

$$u^{(n)(j)}(y,z,t) = \varepsilon_{nj} f_{nj}(y-y_n, z-z_n, t-t_n) \qquad (3.30)$$

## Methodology

where $f_{nj}$ is the shape function of the spot and $\varepsilon_{nj}$ is a random step of value +1 or -1. The shape function determines the spatial and temporal characteristic of the generated turbulence. It has a compact support on [-$\sigma_y$, $\sigma_y$; -$\sigma_z$, $\sigma_z$; -$\sigma_t$, $\sigma_t$] with the length-scale $\sigma$ and satisfies a three-dimensional normalization condition. The number of the active turbulent spots $N$ is kept constant on the inlet plane and approximated with $S_n/S_s$, where $S_n$ and $S_s$ are the surface of the inlet plane and the surface of the support of a spot, respectively. The inlet plane is statistically covered with turbulent spots.

We implemented this method in SPARC and have extended it for compressible flows. In order to take into account the compressibility effects, the fluctuation of the static temperature was added as discussed by Bradshaw [12]. Here it is assumed, that the total temperature fluctuation is negligible compared to the static temperature fluctuation. The latter then reads:

$$\frac{T'}{\overline{T}} \approx -(\gamma-1)Ma^2 \frac{u'}{\overline{u}} \qquad (3.31)$$

where $T$ is the temperature and $Ma$ is the local Mach number and $\gamma$ is the adiabatic exponent. Dashes and overbars denote fluctuations and mean values, respectively.

This assumption is valid for boundary layers and wakes at free-stream Mach numbers $Ma<=5.0$ and for jets at Mach numbers $Ma<=1.5$. This condition is fulfilled in the case of the combustion chamber since there the assumption was taken that the Mach number is $Ma <<1$.

The implementation of the method was firstly tested with the case of the spatially decaying isotropic turbulence. In this case the integral length scale of all spots can be the same and the correlation tensor is also simple. For wall bounded flows, such as the inflow into the burner, there must be a space varying length scale depending on the wall normal distance.

Whereby in [55] there was just a qualitative comparison of some inlet conditions for the test case of spatially decaying turbulence, we verified the method quantitatively with the measurement of Kang et al. [60].

The computational domain is $L_x$=10 $m$, $L_y$=$L_z$=3.52 $m$, where $x$ is the streamwise direction, $y$ and $z$ are spanwise directions. The grid size is $N_x$=144, $N_y$=$N_z$=88.

At the inlet the Synthetic Eddy Method is used with data that correspond to the measurement of Kang et al. [60] at $x/M$=20 behind the active grid ($M$ is the grid size). The integral length scale is set to $l$=0.25 $m$ and is used for each spot, the turbulence intensity is $Tu=15.4\%$. The Reynolds number based on the integral length scale is $Re_l=30600$ and the Mach number of the free stream is $Ma=0.035$.

The shape of the spots is described by a tent function as:

$$f(r) = 1 - r/l \qquad (3.32)$$

with

$$r = \sqrt{y^2 + z^2 + (\overline{u}t)^2} \; . \qquad (3.33)$$

In the two spanwise directions periodic conditions are used.

In order to avoid unphysical reflection of the eddies at the outlet plane, the mesh is stretched gradually until the outlet boundary from the point $x/M=55$ to damp the eddies and additionally at the outlet the static pressure condition with non-reflecting layer is used.

For the discretization in space a second order accurate central difference scheme was used and for the integration in time an explicit four-stage Runge-Kutta scheme was applied. The computations were performed using the Monotone Integrated Large Eddy Simulation (MILES) approach [10]. By this method the SGS structure modelling is based on an artificial dissipation approach in our computations. For the calculation the Matrix artificial dissipation scheme (Swanson and Turkel [134]) based on the non-linear SLIP method (Tatsumi et al. [135]) is used.

The computational domain was initialised with the mean velocity. The time averaging process was started after the first spots have left the last measuring place ($x/M=48$) approximately by four times of the integral length scale. After the time averaging the statistical sample was further increased by averaging over the $x$-planes (homogeneous directions).

Figure 3.12 shows the evolution of the turbulence intensity of the resolved scales downstream the inlet, which corresponds to $x/M=20$ in the measurement. The result compares well with the experimental data.

# Methodology

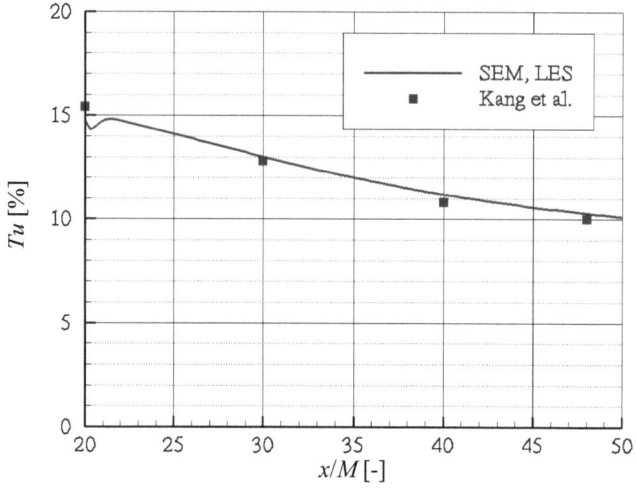

**Figure 3.12:** Evolution of the turbulence intensity of the spatially decaying isotropic turbulence

Figure 3.13 shows by means of iso-surfaces of the $Q$-criterion [48] the structures of the realistic vortices, where $Q$ is defined as:

$$Q = \frac{1}{2}\left(\Omega_{ij}\Omega_{ij} - S_{ij}S_{ij}\right) \qquad (3.34)$$

with

$$\Omega_{ij} = \frac{1}{2}\left(\frac{\partial u_i}{\partial x_j} - \frac{\partial u_j}{\partial x_i}\right) \text{ and } S_{ij} = \frac{1}{2}\left(\frac{\partial u_i}{\partial x_j} + \frac{\partial u_j}{\partial x_i}\right). \qquad (3.35), (3.36)$$

$Q$ is the second invariant of the velocity gradient tensor and represents the local balance between shear strain rate and vorticity magnitude. The $Q$-criterion allows visualization of coherent vortex structures. The value was chosen to visualise the most characteristics structures and therefore it is varying in the present work from case to case depending on the Reynolds number of the given turbulent flow.

Based on these results it is believed that this inlet condition can be successfully used for many other investigations, for example for the simulation of the flow around a turbine blade, where the laminar-turbulent transition depending on the free-stream turbulence intensity is investigated.

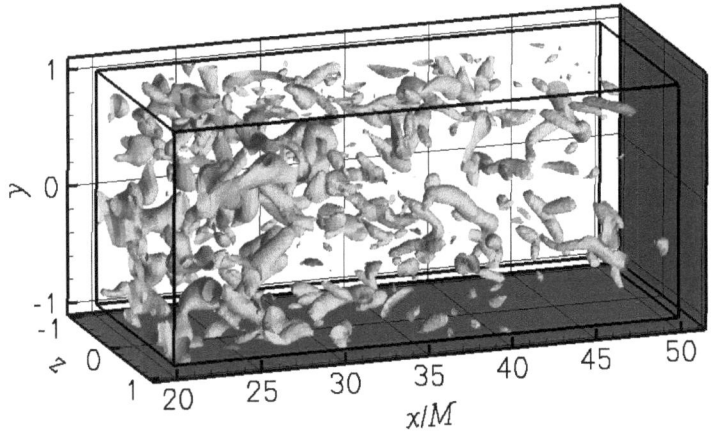

**Figure 3.13:** Iso-surfaces of the $Q$-criterion at the level of 400 $s^{-2}$

In the evolution of the turbulence intensity in Figure 3.12 a small drop can be observed at the first cell, but through the next four cells the flow statistic recovers the physical value. This phenomenon can be seen also in the work of Jarrin *et al.* [55]. Further investigations showed that this fall back and recovering region is strongly depending on the grid resolution. It was found that if the grid is finer the decaying of turbulence is slower.

In order to combine SEM with non-reflecting layer, the spots are generated by source terms at the end of the sponge layer. If the fluctuations are generated at the inlet, the non-reflecting layer damps uncontrollable a part of them.

Furthermore the method was extended to prescribe spots with different length scales placed at different regions of the energy spectrum. Spots larger than the integral length scale are generated based on the $E(\kappa) \sim \kappa$ energy relation in the large eddy range and the smaller spots based on the $E(\kappa) \sim \kappa^{-3/5}$ relation in the universal equilibrium range. This improved the dependency of the decaying from the grid size but further investigation is needed.

The SEM combined with the pulsating mass flow rate inlet condition was tested on the single resonator. The results are described in Section 4.4.1.

## 4 Single Resonator: Combustion Chamber

Former experimental investigations showed that the combustion chamber has specific impact on the stability of the overall system. As first approximation, if the components upstream to the combustion chamber are decoupled by the pressure loss of the coupling element (e.g. burner), the only vibratory component is the combustion chamber, and the system can be treated as a single resonator.

The aim of the investigations of the single resonator is to identify the main damping mechanisms and estimate their effect on the stability of the system.

In Figure 4.1 the sketch of the experimental setup is shown. In the experiments the transfer function of the combustion chamber was calculated from the input signal measured with the hot-wire probe 1 and from the output signal measured with the hot-wire probe 2 [4]. An alternative output signal was the pressure measured with a microphone probe at the middle of the side wall in the combustion chamber [14].

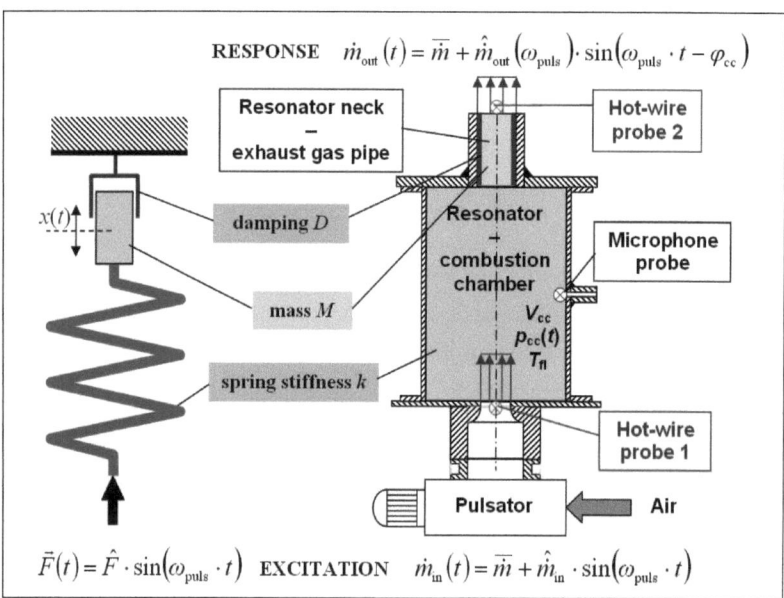

**Figure 4.1:** The sketch of the test rig and the analogy of the mass-spring-damper system and the combustion chamber as Helmholtz resonator

In order to compute the resonance characteristics of the system a series of LESs at discrete forcing frequencies had to be completed. These were taken for a basic configuration corresponding to the experiments, for variation of the geometry of the resonator neck and for variation of the fluid temperature (Table 4.1). The compressible flow in the chamber of the basic geometry was simulated for five different frequencies in the vicinity of the resonance frequency (Case II). Three further LESs were taken for the variation of the geometry (Case III) and other three LESs run for the variation of the fluid temperature (Case IV). Furthermore, one LES was performed without excitation to test the boundary conditions (Case I). In this case the computed velocity profiles were verified with measurements at the vicinity of the nozzle in eight cross sections. The results of Case II showed that the wall roughness in the resonator neck has a significant effect on the damping. In Case V the DEM was used to simulate surface roughness.

In each case the mean mass flow rate was $\bar{m}$ =0.017 $kg/s$, the diameter and length of the chamber were $d_{cc}$=0.3 $m$ and $l_{cc}$=0.5 $m$, respectively and the diameter of the exhaust gas pipe was $d_{egp}$=0.08 $m$. The rate of pulsation was 25% for the Cases II-V.

| Case | $T(K)$ | $l_{egp}(m)$ | $f_{ex}(Hz)$ | wall BC |
|---|---|---|---|---|
| I | 298 | 0.2 | 0 | smooth |
| II | 298 | 0.2 | 37, 39, 40, 41, 42 | smooth |
| III | 298 | 0.1 | 48, 51, 54 | smooth |
| IV | 600 | 0.2 | 48, 54, 60 | smooth |
| V | 298 | 0.2 | 38, 40, 42 | rough |

**Table 4.1:** Parameter set of the simulations

In the case of the combustor the compressible Navier-Stokes equations are solved. The spatial discretization is a second-order accurate central difference formulation. The temporal integration is carried out with a second-order accurate implicit dual-time stepping scheme. For the inner iterations the 5-stage Runge-Kutta scheme was used. The time step was $\Delta t$=2·10$^{-5}$ $s$. This was a compromise in order to resolve the turbulent scales and compute the pulsation cycles within the permitted time. The Smagorinsky-Lilly model was chosen as subgrid-scale model [70]. Later investigations with MILES approach, dynamic Smagorinsky model show no significant difference in the results. This proofs that the mesh is sufficiently fine in the regions which are responsible for the damping of the pulsation, thus the modelling of the SGS structures has a minor influence there.

## 4.1 Computational Domain and Boundary Conditions

If the flow in the combustion chamber and the resonator neck has to be simulated (grey area in Figure 4.2) attention should be paid to some difficulties by the definition of the boundary conditions.

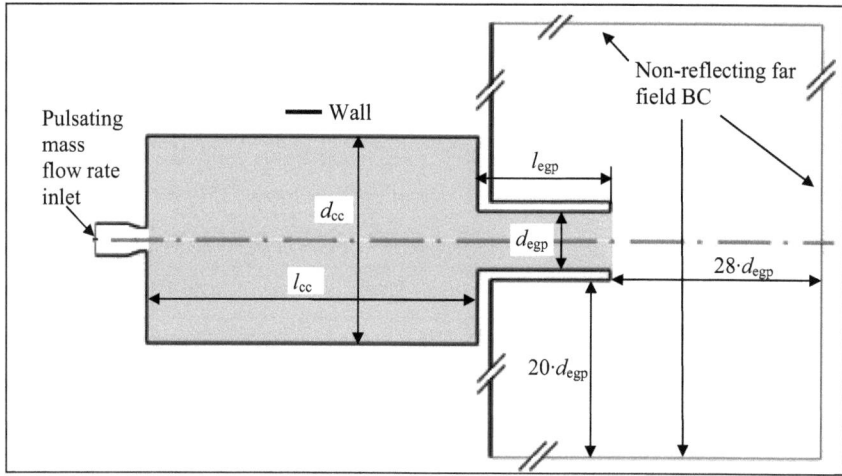

**Figure 4.2:** Sketch of the computational domain and boundary conditions

Even though the geometry of the chamber is axisymmetric no symmetry or periodic condition could be used because the vortices in the flow are three-dimensional and they are mostly on the symmetry axis of the chamber. In the present simulations an O-type grid is used to avoid singularity at the symmetry axis.

At the inflow boundary the fluctuation components must be prescribed for a LES. Furthermore, the boundary should not produce unphysical reflections, if the pressure fluctuations, which move in the chamber back and forth, go through the inlet. A conventional boundary condition can reflect up to 60% of the incident waves back into the flow area. One can avoid these reflections only by the use of a non-reflecting boundary condition. If the inlet would be set at the boundary of the grey area, this problem can be solved hardly. In the experimental investigation a nozzle was used at the inflow into the chamber. The pressure drop of the nozzle ensures that the gas volume in the test rig components upstream of the combustion chamber does not affect the oscillation response of the resonator. It was decided to use this nozzle in the computation also. Although the additional volume of the nozzle increases the number of computing cells, a non-reflecting boundary condition is no more necessary. In addition, the fluctuation components at the inlet can be neglected, since the nozzle decreases strongly the turbulence level downstream.

The inflow condition prescribes the pulsating mass flow rate based on the velocity profile measured by the experiment. The pulsation was achieved by scaling the velocity profile in time.

The definition of the outflow conditions at the end of the exhaust pipe is particularly difficult. The resolved eddies can produce local backflow in this cross section occasionally. In particular, by excitation frequencies in the proximity of the resonant frequency there is a temporal backflow through the whole cross section, which has been observed by the experimental investigations as well.

The change of the direction of the flow changes the mathematical character of the set of equations. For compressible subsonic flow four boundary values must be given at the inlet and one must be extrapolated from the flow area. At the outlet one must give one boundary value and extrapolate four others. Since these values are a function of the space and time, their determination from the measurement is impossible. Further the reflection of the waves must be avoided also at the outlet. For these reasons the outflow boundary was set not at the end of the exhaust gas pipe, but in the far field. In order to damp the waves in direction to the outlet boundary, a buffer layer with mesh stretching is used.

At the solid surfaces the no-slip boundary condition and an adiabatic wall were imposed. For the first grid point $y^+<1$ was obtained, the turbulence effect of the wall was modelled with the van Driest type damping function. The geometry of the computational domain and the boundary conditions are shown in Figure 4.2. The entire computational domain contains about $4.3 \cdot 10^6$ grid points in 111 blocks. A coarsened mesh is shown in Figure 4.3.

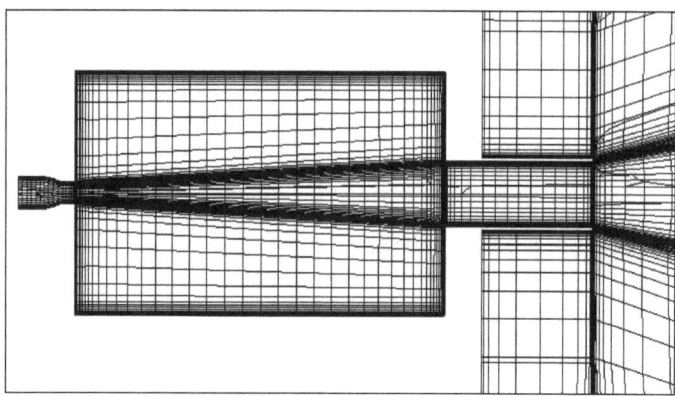

**Figure 4.3:** Third finest mesh extracted to the symmetry plane (distortions were caused by the extraction in Tecplot)

## 4.2 Flow Properties in the Combustion Chamber

The flow in the resonator is highly turbulent. The Reynolds number based on the mean bulk velocity is about $3.7 \cdot 10^4$ in the nozzle and in the exhaust gas pipe. The shear flow downstream of the nozzle and the edges of the resonator neck (separation bubbles) are the major vortex shedding zones (Figure 4.4).

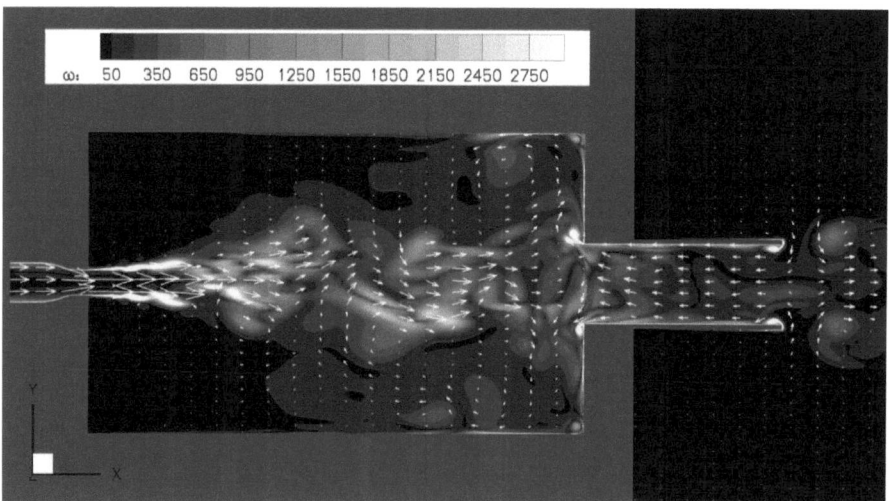

**Figure 4.4:** Instantaneous vortex structures and velocity vectors in the symmetry plane

The compressibility plays a huge role at the resonance frequency in particular. The gas volume in the chamber acts as a spring, therefore it is possible, that gas flows in the chamber from the nozzle as well as through the resonator neck in the resonator at the same time (Figure 4.4).

By the formulation of the reduced physical model it was assumed, that the pressure is in phase and location-independent in the resonator. In order to prove this assumption the pressure distribution was investigated during the pulsation. In Figure 4.5 four uniformly distributed times are marked during a pulsation period for Case II in the vicinity of the resonance frequency. In this figure the inlet and outlet mass flow rate and the pressure in the middle of the side wall of the chamber are depicted. At these times the pressure distribution along the symmetry axis and on the side wall of the chamber is plotted in Figure 4.6. Aside from the pressure fluctuation inside the jet it can be clearly seen that the pressure has the same level in each region of the chamber.

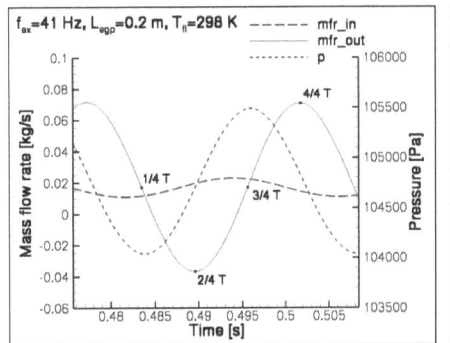 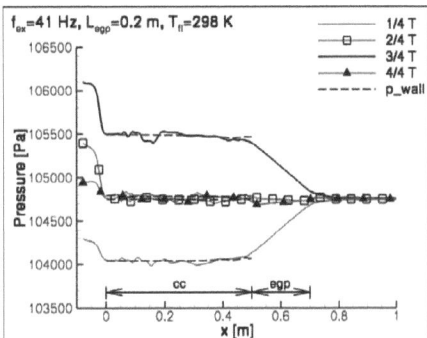

**Figure 4.5:** One period of the mass flow rate and pressure signal of Case II

**Figure 4.6:** Pressure at the symmetry axis and at the wall of the combustion chamber at different fractions of a period of pulsation (Case II)

Figure 4.5 illustrates also the 90° phase shift between the input and output mass flow rate and between the output pressure signal and the output mass flow rate, as well.

As mentioned above, in the vicinity of the resonant frequency, there is a backflow through the whole cross section at the end of the exhaust pipe. In this case a strongly oscillating flow exists in the pipe. The exact solution of the Navier-Stokes equations for the flow near an oscillating flat plate (2$^{nd}$ problem of Stokes) gives the boundary layer thickness for the laminar case [122]. This can serve for the turbulent flow in the exhaust gas pipe as a rough estimation. The analytical solution predicts a boundary

## Single Resonator: Combustion Chamber

layer thickness of $\delta=1.7$ mm, which is depending on the kinematic viscosity and the frequency of oscillation:

$$\delta \propto \sqrt{\frac{\nu}{f}}. \tag{4.1}$$

Despite the relatively coarse resolution (about 20 points covered in the boundary layer), the calculation with about $\delta=2$ mm shows a good approximation. For oscillating laminar flow in channels or in pipes the exact solution of the Navier-Stokes equations can be derived [122]. This shows that the maximum value of speed does not coincide with the axis of the pipe, but occurs near the wall. This so called annular effect was confirmed also by measurements [114], [125]. Since there is a nonzero mean flow rate in the case of the combustion chamber, this effect can be observed especially during the back flow phase. The results of the simulation agree quite well with the analytical solution.

In order to save computational time wall models were extended for LES by e.g. Schumann [123]. The measurement of oscillating flows shows, that the logarithmic region of the mean velocity profile decay in about 20% of an oscillation period already at an excitation frequency of $f_{ex}=0.1$ Hz [56]. The surveys of Tsuji and Morikawa [142] also show that when the free stream is strongly accelerated and decelerated, departures from the universal logarithmic law occur. Scotti and Piomelli investigated a pulsating turbulent channel by means of DNS and LES [124]. The pulsating flow can be described by three parameters: mean friction velocity $u_\tau$, forcing frequency in wall units $\omega^+=\omega\nu/u_\tau^2$ and the ratio between oscillating and mean centreline velocity $a_{uc}$. In [124] current-dominated flows ($a_{uc}<1$) were investigated and it was shown that depending on the excitation frequency the logarithmic layer is present only through part of the cycle or shifted upwards and downwards.

In Case II at the resonance these parameters are $u_\tau=0.635$, $\omega^+=0.0092$ and $a_{uc}=9.8$ i.e. the oscillating part is dominating (wave dominating flow) with a low frequency pulsation. Thus the flow can be characterized as an oscillating flow with relaminarization and laminar-to-turbulent transition. The velocity profile in the exhaust pipe was examined at different places and for different phases. The distribution of the velocity normal to the wall at the half length of the exhaust gas pipe at different phases of the pulsation cycle is plotted in Figure 4.7 (see Figure 4.5 for the time phases). The results show that the velocity profile of the oscillatory boundary layer deviates predominantly from the logarithmic law of the wall. This supports the necessity of fine resolution of the boundary layer up to the wall in the case of the combustion chamber. The use of the logarithmic wall function in pulsating flows was investigated also in later work with similar results [100].

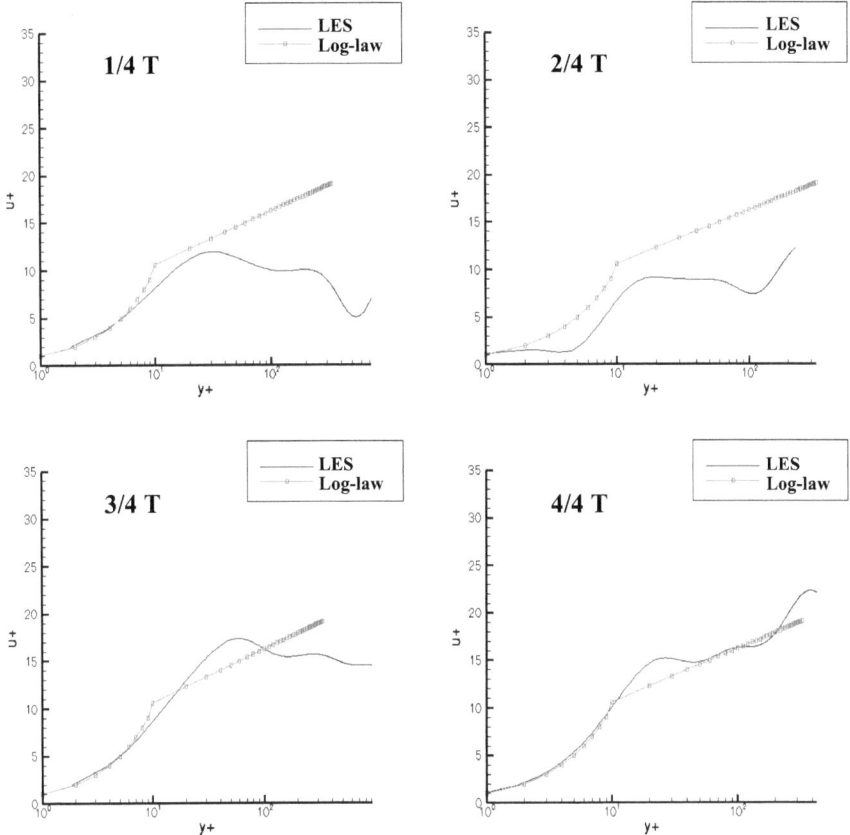

**Figure 4.7:** Velocity distributions at the half length of the exhaust gas pipe at different fractions of the period of pulsation

## 4.3 Damping Mechanisms

In this chapter some mechanisms referred in the literature are investigated, which could damp the pulsation in the system.

In the case of the combustion chamber the mean pressure is 1 *bar* and it is pulsating by approx. 1500 *Pa*, furthermore the Mach number is quite low. The bulk viscosity, which can have a significant influence on sound propagation and shock-wave structures, has in this case no effect on the pulsation [144]. As the time scale of the pulsation is much smaller than the time scale of heat transfer, the compression and expansion process in the chamber can be treated as adiabatic thus the losses through the pressure-volume work can be neglected.

It was expected, that the oscillating boundary layer plays the major role in the damping of the pulsation. In order to compare the effect of friction in the steady mean and the pulsating flow the dissipation function was calculated [141]. The dissipation function appears in the energy equation and it shows how much mechanical energy is dissipated through viscous stresses. It is defined as:

$$\Phi = \mathrm{div}(\mathbf{T}\vec{v}) - \vec{v}\mathrm{Div}\mathbf{T}. \qquad (4.2)$$

Applying the Stokes' law it can be written as:

$$\begin{aligned}\frac{\Phi}{\mu} =\ & 2\left[\left(\frac{\partial u}{\partial x}\right)^2 + \left(\frac{\partial v}{\partial y}\right)^2 + \left(\frac{\partial w}{\partial z}\right)^2\right] + \left(\frac{\partial v}{\partial x} + \frac{\partial u}{\partial y}\right)^2 + \left(\frac{\partial w}{\partial y} + \frac{\partial v}{\partial z}\right)^2 \\ & + \left(\frac{\partial u}{\partial z} + \frac{\partial w}{\partial x}\right)^2 - \frac{2}{3}\left(\frac{\partial u}{\partial x} + \frac{\partial v}{\partial y} + \frac{\partial w}{\partial z}\right)^2\end{aligned}. \qquad (4.3)$$

The dissipation function was calculated for Case I and for Case II at $f_{ex}$=41 $Hz$. The iso-surfaces at 3000 $J/sm^3$ are plotted in Figure 4.8 and in Figure 4.9.

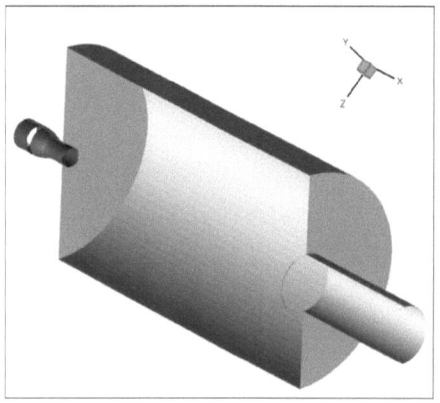
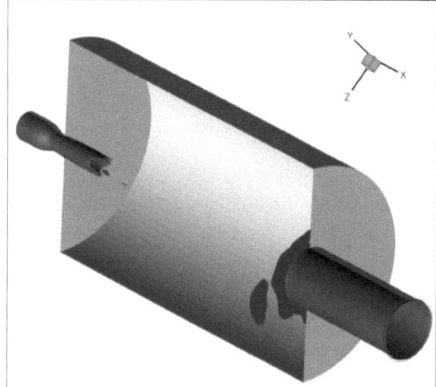

**Figure 4.8:** Dissipation of the flow without excitation

**Figure 4.9:** Over one period averaged dissipation of the flow with excitation near the resonant frequency

In the case of the pulsating flow (Figure 4.9) the dissipation function was calculated in 40 discrete points in time of a period of the pulsation and the average of them was taken. The figures show clearly that in the case of pulsation an additional region of high dissipation appears in the resonator neck.

In Figure 4.10 the distribution of the dissipation function is plotted in radial direction at the half length of the exhaust gas pipe. It shows that in the vicinity of the wall the dissipation increases from approx. 90 $J/sm^3$ of the steady mean flow to approx. 6000 $J/sm^3$ of the pulsating flow.

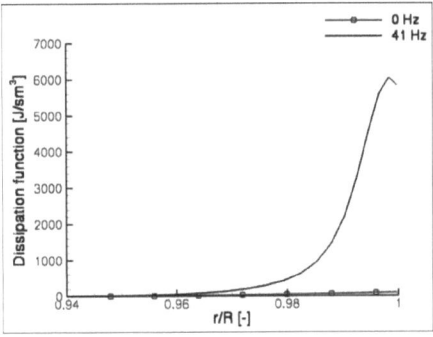 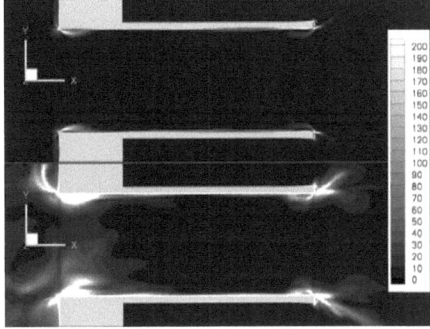

**Figure 4.10:** Distribution of the dissipation at the half length of the exhaust gas pipe in radial direction

**Figure 4.11:** Dissipation in the region of the resonator neck; top: steady mean flow, bottom: pulsating flow

In Figure 4.11 the distribution of the dissipation in the exhaust pipe is compared between the steady mean flow and the pulsating flow. It shows that beyond the oscillating boundary layer the separation bubbles at the ends of the exhaust pipe play an important role.

The generation of turbulence by the pulsation is very demanding to investigate. It needs phase averaging over many cycles of the pulsation which was not available within the time limitation of the project (one cycle was resolved with approx. 2500 time step).

However, to see the difference in the generation of turbulence by the steady mean flow and the pulsating flow the $Q$-criterion was calculated and compared. In Figure 4.12 and Figure 4.13 the iso-surfaces of the $Q$-criterion at the level of $2 \cdot 10^5 \ s^{-2}$ for Case I and Case II at $f_{ex}=41\ Hz$ are plotted. In the case of the non-pulsating flow fine turbulent structures in the jet can be observed which can be found also in the exhaust gas pipe (Figure 4.12).

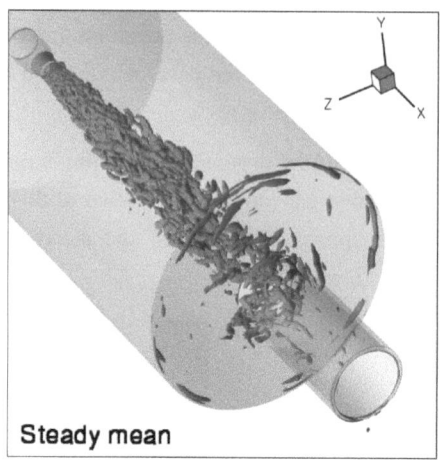

**Figure 4.12:** Iso-surfaces of the $Q$-criterion in the steady mean flow at the level of $2 \cdot 10^5 \ s^{-2}$

In contrast the structures in the pulsating jet are much coarser but fewer and only some structures reach the exhaust gas pipe (Figure 4.13). The pressure is rising in the combustion chamber during the acceleration phase of the pulsating jet which can act on the eddy structures. Furthermore, as the accelerated fluid reaches the exhaust pipe it will be stopped by the reverted flow in the resonator neck. Thus a temporary stagnation point exists there which can be seen in Figure 4.4. This latter is property of the present configuration. If the chamber length would be increased, and the width decrease to keep the resonant frequency on the same value, or vice versa, the flow pattern and maybe also the resonance characteristics would change.

In Figure 4.13 the ring vortex shedding (separation bubbles) at the ends of the exhaust gas pipe can be detected clearly as typical flow patterns of the pulsating flow. These structures increase the damping which was shown also by the dissipation function in Figure 4.11. In order to estimate the dissipation caused by the turbulence in other regions the phase averaging is indispensable.

# Single Resonator: Combustion Chamber

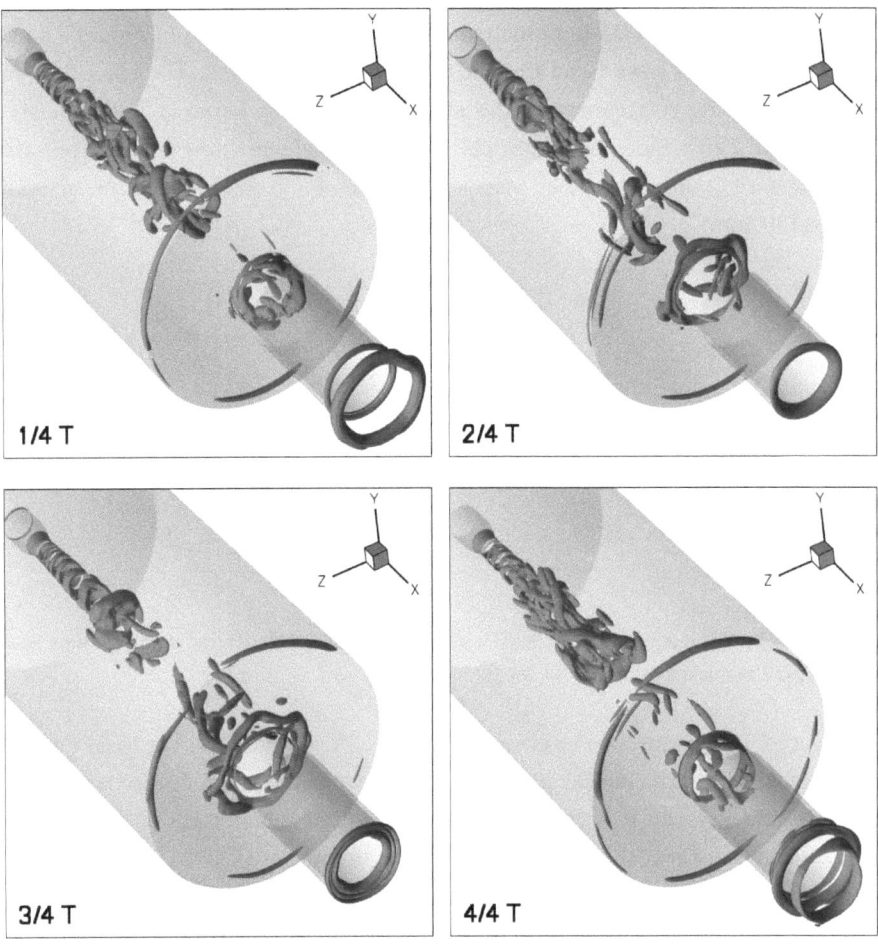

**Figure 4.13:** Iso-surfaces of the $Q$-Criterion in the pulsating flow at the level of $2 \cdot 10^5 \ s^{-2}$

It was mentioned above that temporarily there is a stagnation point between the jet and the reversed flow in the resonator neck. In Figure 4.4 a further temporary stagnation point can be observed at the outer end of the exhaust gas pipe, too. During the pulsation cycle it can be observed, that at large amplitudes, if reversed flow in the exhaust pipe occurs, the ambient air is streaming sideways into the pipe while the core of the outer jet preserves the original flow direction. This can be better observed in Figure 4.16 by means of the temperature difference. Based on this observation it was assumed, that the pulsating flow cause a higher entrainment which include a higher impulse transfer to the ambient air, thus the damping of the pulsation increases. To prove the assumption the mass flow rate was computed at the $x$ plane approx. $0.5\ m$ streamwise from the exit of the exhaust gas pipe. In Figure 4.14 the mass flow rates of Case I and Case II at $f_{ex}=41\ Hz$ are plotted. The mass flow rate increased by the entrainment of the steady jet of Case I (mfr_x_1300mm_st) is approx. $0.037\ kg/s$, more than the double of the original mass flow rate. The pulsating jet, contrarily, increases the mass flow rate to $0.065\ kg/s$ of mean value (mfr_x1300mm).

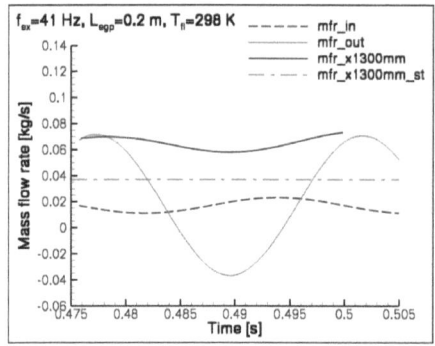
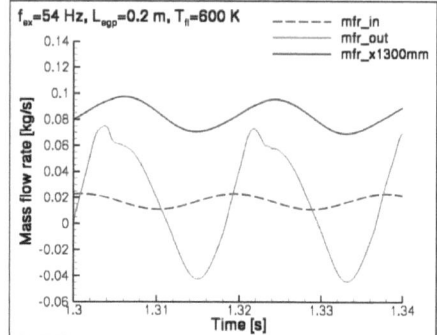

**Figure 4.14:** Entrainment of the steady jet of Case I and the pulsating jet of Case II

**Figure 4.15:** Entrainment of the pulsating jet of Case IV

In Figure 4.15 the same can be observed for Case IV at $f_{ex}=54\ Hz$. The mass flow rate at the inlet is the same but because of the higher temperature the density is lower and therefore the velocity is higher. This can increase the entrainment of the steady jet also, which is not available from computation, unfortunately.

Hudson [45] measured the thrust on a piston driven half-open tube and investigated the flow at the open end of the tube. He describes the flow with four phases: jet formation (14/16 T in Figure 4.16), jet separation from the gas column in the pipe (6/16 T), vortex formation (7/16 T) and laminar inflow (10/16 T). In the case of the combustion chamber the phase of the jet formation takes a longer time because of the nonzero mean flow rate. Hudson investigated theoretically the thrust limitation by the jet and the acoustic radiation. He found, that the acoustic limitation is negligible small and most of the energy is carried away in the jet and the associated vortex formation. He obtained an excellent agreement between theory and experiment.

Based on these the improvement of the entrainment arises from the suction of fluid sideways, ring vortex shedding [99] and the increasing of turbulence intensity. The first one depends on the amplitude of the pulsation. It appears only if reversion of the flow happens.

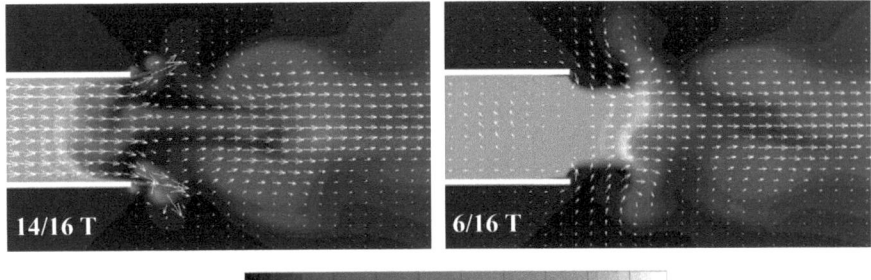

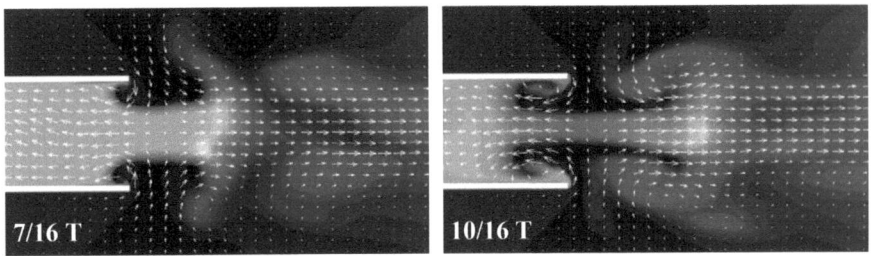

**Figure 4.16:** Distribution of the temperature at different fractions of the pulsation cycle
(Case III at $f_{ex}$=54 $Hz$)

In the previous subsection three parameters of the pulsating/oscillating flow were reviewed. In order to make an estimation of the damping of the pulsation and the resistance of the system further variables as the current Reynolds number and the oscillatory-flow (wave) boundary layer Reynolds number have to be taken into account. Lodahl *et al.* [71] showed that the overall average mean wall shear stress as well as the oscillating wall shear stress is strongly depending on the flow regimes. If the flow is wave-dominated the averaged mean shear stress can increase significantly if the oscillatory-flow component is turbulent or even decrease if the oscillatory-flow component is laminar. Manna *et al.* [91] reached an overall space and time averaged resistance reduction up to 27%. The oscillatory component of the wall shear stress can be higher of several orders, if both the oscillatory boundary layer and the main stream are fully turbulent.

## 4.4 Comparison of the Results with Experimental Data

In this section the resonance characteristics of the combustion chamber obtained from experiments and computations are compared by means of the amplitude and phase transfer functions. The amplitude ratio of the mass flow rates is defined as:

$$A = \frac{\hat{\dot{m}}_{out}}{\hat{\dot{m}}_{in}}. \qquad (4.4)$$

Figure 4.17 illustrates a typical evolution of the mass flow rate at the outlet of the exhaust gas pipe to cycle limit. The amplitude ratios were calculated if the cycle limit was reached.

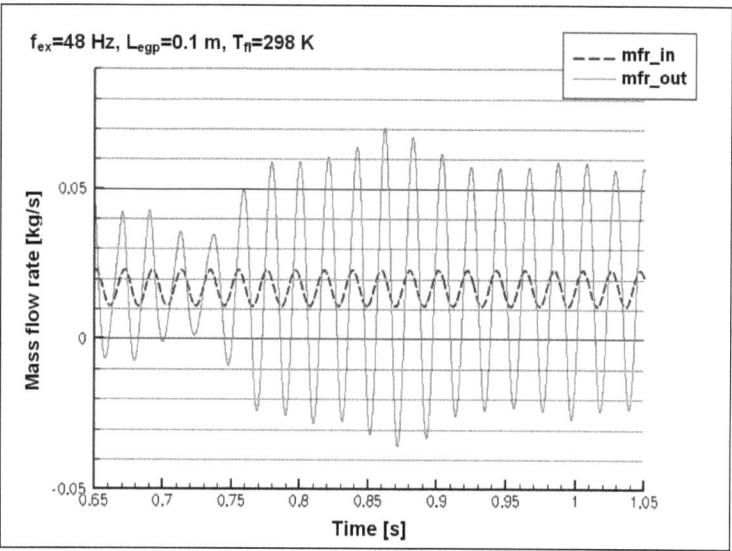

**Figure 4.17:** Mass flow rate evolution to cycle limit

### 4.4.1 Case II: Basic Configuration

In Figures 4.18 and 4.19 two experimental data sets with the analytical model and the results of the computation are exhibited. In one case of the experiments the exhaust pipe was manufactured from a turned steel tube, in the other case the tube was polished. The LES data compare more favourably with the experimental data of polished tube, because the wall in the simulation was aerodynamically smooth, just like the polished resonator neck. The computation predicts the damping factor quite well; the deviation is about 7%. If the results of the measurement of the turned steel tube are compared with the simulation, the deviation is about 40%.

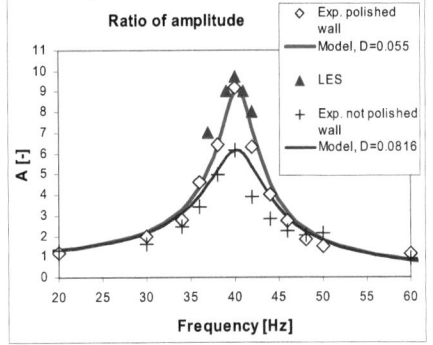 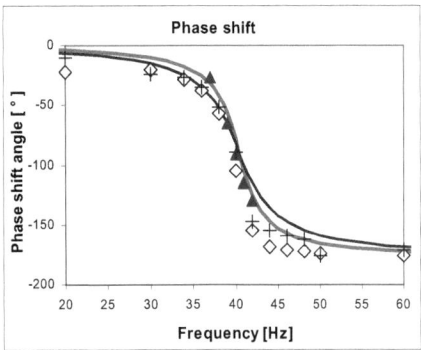

**Figure 4.18:** Amplitude response of the combustion chamber

**Figure 4.19:** Phase transfer function of the combustion chamber

These results shed light on the strong influence of the surface quality of the wall in the exhaust gas pipe on the damping, which was not investigated systematically before.

For this basic configuration the mass flow rate with precursor simulation and the SEM was tested to investigate the effect of the inlet turbulence on the resonance characteristics of the resonator. In the case of the precursor simulation a precursor domain was coupled to the inlet of the domain shown in Figure 4.2 and the mesh in the inlet region was refined. The length was set to $3 \cdot d_{inlet}$ to cover safely the space correlations. The mesh had approx. $0.7 \cdot 10^6$ extra CVs compared to the original one. For the case with SEM the mesh was refined only at the inlet region, hence the mesh had approx. $0.2 \cdot 10^6$ extra CVs compared to the original one.

The precursor simulation is integrated in the main simulation, thus the turbulent flow could develop in the precursor region simultaneously to the flow in the chamber. Despite this the extra computational time is meaningful till the fully turbulent state in the precursor region is reached. Furthermore the flow in the chamber becomes much faster fully turbulent due to the strong shear layer of the jet. From the channel flow simulation it was known that an initialization with random fluctuation cannot produce a faster evolution of turbulence. The random fluctuations were dissipated very fast, a laminar state was reached and later a physical transition has happened.

## Single Resonator: Combustion Chamber

SEM produces, however, fluctuations immediately from the beginning of the computation and needs additionally much less extra CVs.

The result of the investigation with SEM at $f_{ex}=40\ Hz$ shows that in the case of the combustion chamber the inlet turbulence has no remarkable effect on the resonance characteristics. This can be explained with the dominance of the turbulence generated by the jet in the chamber (Figure 4.20).

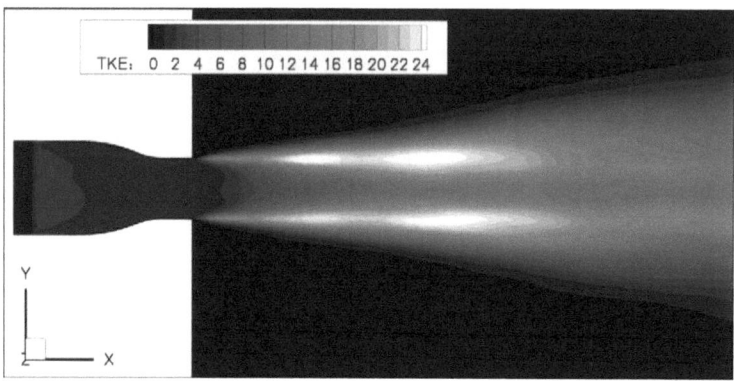

**Figure 4.20:** Distribution of the turbulent kinetic energy in the symmetry plane (Case I with SEM)

### 4.4.2 Case III: Variation of the Geometry

In the case with shorter resonator neck the results of the simulations compare very well with the experimental data (Figure 4.21 and 4.22).

It must be noticed that despite the shorter exhaust gas pipe, i.e. smaller area of dissipation (see Section 4.3), the damping is higher than in Case II. This is due to the higher resonance frequency, which causes a thinner boundary layer with higher velocity gradient and shear stress (see Equation (4.1) and Section 4.4.4).

Special feature of this case is that the ring vortices generated at both ends of the exhaust gas pipe are not damped and the relaminarization of the flow is weaker because of the shortness of the pipe (Figure 4.23).

Single Resonator: Combustion Chamber

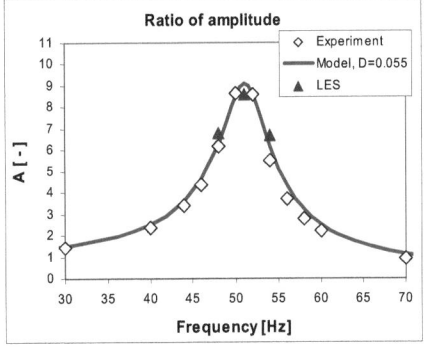 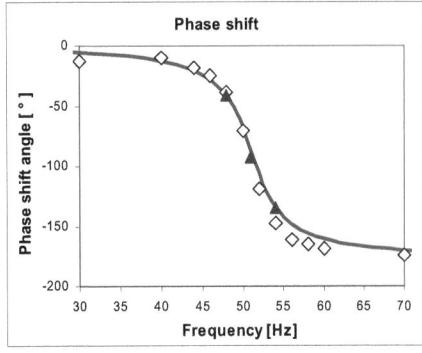

**Figure 4.21:** Amplitude response of the combustion chamber

**Figure 4.22:** Phase transfer function of the combustion chamber

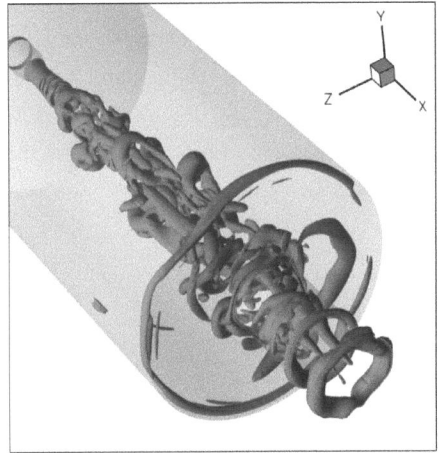

**Figure 4.23:** Iso-surfaces of the $Q$-criterion at the level of $2 \cdot 10^5 \ s^{-2}$

### 4.4.3 Case IV: Variation of the Fluid Temperature

For the case of higher fluid temperature there is some difficulty to evaluate the system response from the computation with the mass flow rate measured in the proximity of the resonant frequency. In the experiments the pressure signal in the resonator was measured as response signal for this case. This alternating method of response signal measurement is derived in [14] and proved by experiments. The results of pressure measurement were converted on the basis of the equation in [14] into mass flow rate amplitudes, in order to compare the results like in the case of the basic configuration.

Unfortunately, only the resonance characteristics of the resonator neck with rough wall were available from the experiments for this case. Based on the results from the Case II an approx. 40% higher amplitude ratio was expected from the simulations. However, the comparison showed more than 100 % deviations.

During the backflow cold air comes in sideways from the environment (Figure 4.16). Because of the cold air with higher density the mass flow rate signal at the end of the resonator neck has additional peeks (signal Aout/Ainmax in Figure 4.24). This causes the error in the determination of the real amplification. Therefore, the pressure signal was also stored during the simulation at the centre of the side wall of the combustion chamber, at the same place, where it was measured in the experiment. The amplitudes of the normalized pressure oscillation were converted into mass flow rate oscillation (p_amp in Figure 4.24). It is compared with the mass flow rate oscillations determined also from the simulation (Ain/Ainmax and Aout/Ainmax) in Figure 4.24. In this figure the signals are shifted by the mean mass flow rate and they are normalized by the amplitude of the mass flow rate signal at the inlet (Ainmax) for the sake of better display of the amplification.

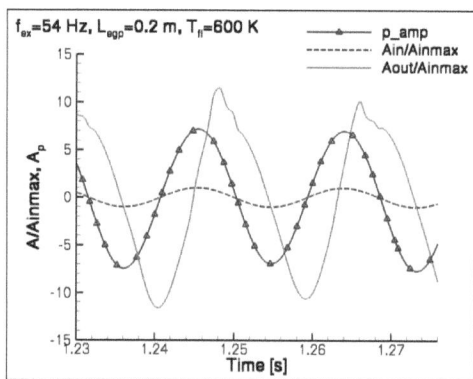

**Figure 4.24:** Amplitude ratios from mass flow rates and pressure signal, respectively

By means of this calculation the difference was reduced to about 40% in the proximity of the resonant frequency. This corresponds to the result of the basic configuration, where the values from the simulation were compared with the measurement of the turned tube. In Figure 4.25 the mass flow rate amplitude ratio determined from the pressure fluctuations are presented. In Figure 4.26 the phase shift is plotted.

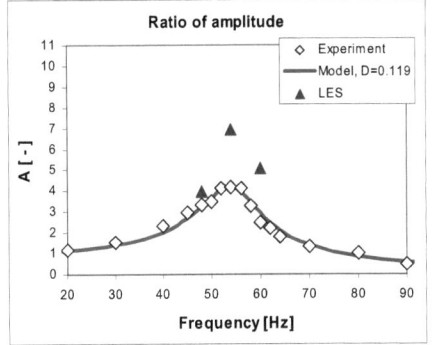

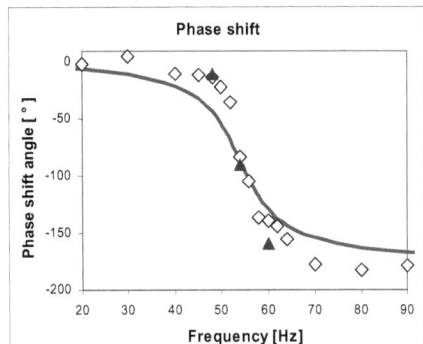

**Figure 4.25:** Amplitude response of the combustion chamber

**Figure 4.26:** Phase transfer function of the combustion chamber

Figure 4.16 shows that at the symmetry axis the temperature is predominantly high during a pulsation cycle while cold flow is sucked sideways from the ambient. A measurement with a hot wire in the centre point of the cross section at the end of the exhaust gas pipe would provide much lower amplitude of the mass flow rate pulsation if a cylindrical velocity distribution with homogeneous temperature is assumed.

### 4.4.4 Case V: Simulation with Roughness

For this numerical investigation the Discrete Element Method was chosen and implemented in SPARC (see Section 3.6.1). In order to compute the resonance characteristics of the system influenced by the surface roughness in the exhaust gas pipe the same methodology was chosen as in the case of the smooth wall. The only information from the experiments was that the two different sets of data were achieved from a measurement with an exhaust gas pipe made of a polished steel tube and of a machined steel tube, respectively. The polished steel tube could be treated as an aerodynamically smooth wall. The roughness height for the machined steel tube was unknown and it was estimated to be in the range of equivalent sand grain roughness $k_S=0.02 \div 8$ $mm$ [150]. Based on the calibration of the cone height with sand grain size by Bühler [17] the height of the cones was then estimated to $h_c=0.002 \div 0.2$ $mm$. Therefore, a few simulations were carried out with different heights of the virtual cones, each with an excitation frequency of $f_{ex}$=40 Hz (approximately the resonance frequency of the system). The result closest to the experiment with the machined tube was taken and two other excitation frequencies were computed.

In order to understand the importance of surface roughness in the case of the combustion chamber one has to investigate the pulsating flow in the exhaust gas pipe. The previous investigations showed that the pulsation of the flow produces an additional dissipation of mechanical energy in the resonator neck. For oscillating flow in channels or in pipes the exact solution of the Navier-Stokes equations can be derived [122]. This shows that the maximum value of velocity does not coincide with the axis of the pipe, but occurs near the wall (annular effect, see Section 4.2). Since there is a nonzero mean flow rate in the case of the combustion chamber, this effect is slightly asymmetric. In Figure 4.27 a) the velocity distributions normal to the wall are plotted at the half length of the exhaust gas pipe. The streamwise components are normalized by the velocity at the pipe axis ($r$=0), which are approx. $U_{r=0,\text{non-puls}}$=3.4 $m/s$ and $U_{r=0,\text{puls}}$=12.5 $m/s$. The velocity profile for the pulsating flow is taken for this plot at the time of maximum outflow and from the pulsation near to the resonance frequency.

The most obvious difference is that the profile of the pulsating flow has a local maximum near to the wall which is also predicted by the analytical solution. The consequence is a much higher velocity gradient and shear stress on the wall. In Figure 4.27 b) the velocity profiles of the pulsating flow are plotted during a period (see Figure 4.5 for the time phases). The plots demonstrate the continuous presence of the local maximum of the velocity near the wall. At the last quarter of the period (3/4 T) e.g. the value of the local maximum exceeds the maximum value of the non-pulsating flow, which is, however, at the axis of the pipe. The high gradient of the velocity explains the sensitivity of the flow to the surface roughness. If the roughness magnitude increases slightly, a region of much higher velocities is reached and so the drag force increases significantly.

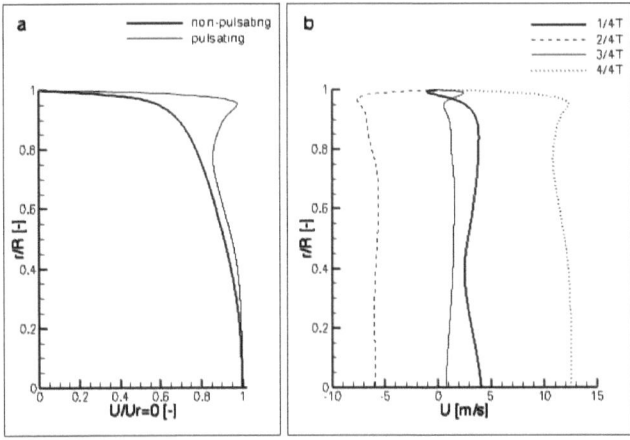

**Figure 4.27:** Velocity distributions at the half length of the exhaust gas pipe. a) Normalized U-profile of the non-pulsating and of the pulsating flow. b) U-profiles of the pulsating flow at different fractions of the period of pulsation

The exact value of the roughness in the machined steel tube was not available; therefore more calculations were carried out with different virtual cone heights. The results of $h_c$=0.5 mm are closest to the experimental data of the exhaust gas pipe with rough wall. In Figure 4.28 the frequency responses of the smooth and rough configuration are indicated.

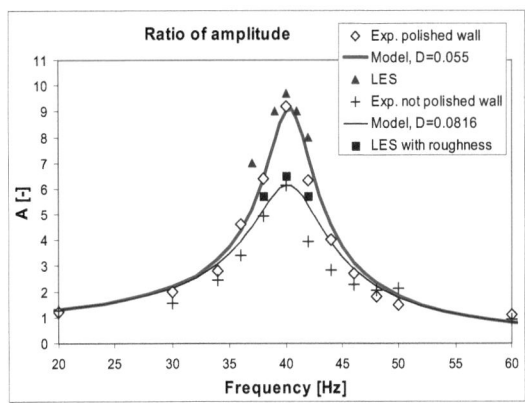

**Figure 4.28:** Frequency response curve of the combustion chamber

In [17] the calibration of the virtual cones' height with the sand grain roughness is described. A further work is needed to calibrate the cones' height with the roughness factor mostly used in practice.

# 5 Coupled Resonators: Burner and Combustion Chamber

The model of the single Helmholtz resonator describes combustion systems sufficiently precise only in a first approximation, since real combustion systems in general have more vibratory gas volumes in addition to the combustion chamber (mixing device, air/fuel supply, burner plenum and exhaust gas system). The linking of these vibratory subsystems results in a significantly more complex vibration behaviour of the overall system compared to the single combustion chamber. To get closer to real combustion systems the model of the single Helmholtz resonator must be extended to describe more resonators coupled to each other.

For modelling a coupled system the burner plenum was added upstream to the combustion chamber. The reduced physical model was extended for the coupled system of burner and combustion chamber [116]. In order to prove the prediction of the model for the coupled system different geometric parameters (burner volume, resonator geometry) and operating parameters (mean mass flow rate) were varied in the experimental part. In each case the flow was non-reacting.

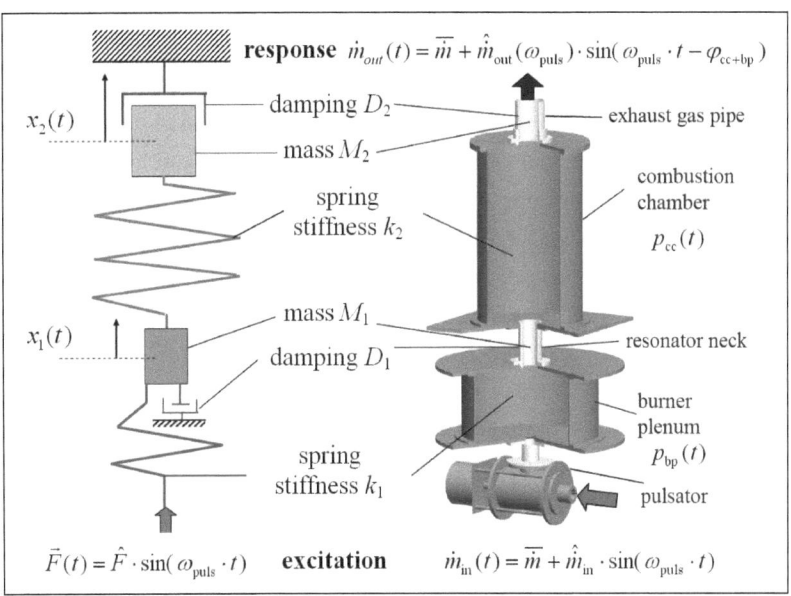

**Figure 5.1:** Coupled Helmholtz-resonators and oscillating masses connected with springs and damping elements

By means of numerical simulation and the physical model the resonance characteristics of a combustion system can be predicted already during the design phase. The main goal of the numerical investigation is to predict the damping coefficient of the system which is an important input for the physical model. In order to provide an insight into the flow mechanics inside the system LESs were carried out. The solution of the fully compressible Navier-Stokes equations was essential to capture the physical response of the pulsation amplification, which is mainly the compressibility of the gas volume in the chamber. Viscous effects play a crucial role in the oscillating boundary layer in the neck of the Helmholtz resonator and, hereby, in the damping of the pulsation. The pulsation and the high shear in the resonator neck produce highly anisotropic swirling flow. Therefore it is unlikely that an URANS simulation can render such flow reliably.

The results of the investigations showed that LES can predict accurately the resonant characteristics of the single resonator and the damping, respectively. In this section the ability of LES predicting the resonant characteristics of the coupled system will be investigated.

The sketch of the experimental setup and the analogy of a mass-spring-damper system are shown in Figure 5.1. On the right side of the figure the system of two coupled Helmholtz-resonators is indicated. The mean volume flow rate is partially pulsated by the pulsator unit. The pulsating flow passes through the burner plenum, reaches the combustion chamber through the resonator neck and leaves the system at the end of the exhaust gas pipe. In the experiments the pulsating mass flow rate was measured at the entrance of the system and at the exit cross section of the exhaust gas pipe to determine the resonance behaviour of the coupled resonator system. Figure 5.2 exhibits the test rig of the coupled Helmholtz resonators.

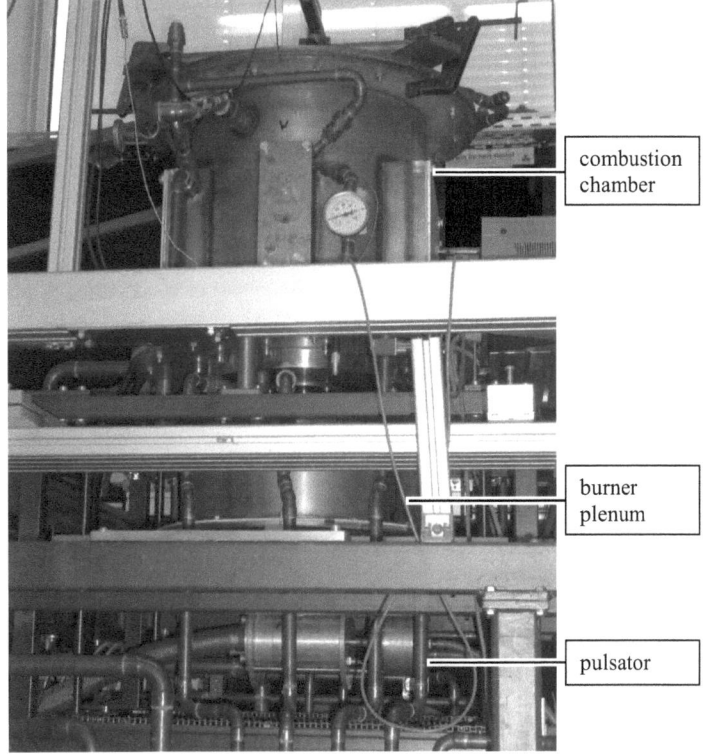

**Figure 5.2:** Test rig of the coupled Helmholtz-resonators

## 5.1 Computational Domain and Boundary Conditions

The geometry of the configuration chosen for the numerical investigation is illustrated in Figure 5.3. The observation windows and the inserted baffle plates increased the complexity of the geometry and hence the generation of the mesh significantly. There were baffle plates placed in the burner plenum and in the combustion chamber to avoid the jet of the nozzle and of the resonator neck to flow directly through the system, furthermore to achieve a homogeneous distribution of the velocity in the cross-section of the measuring point at the end of the exhaust gas pipe.

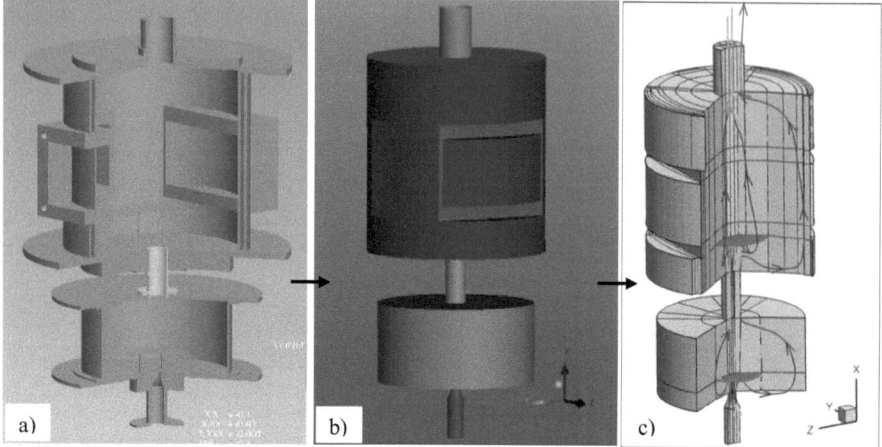

**Figure 5.3:** a) Geometry of the test rig, b) definition of the flow domain, c) 3D block-structured mesh

The definition of the computational domain and the boundary conditions is akin to the case of the single Helmholtz-resonator. The inlet surface was placed upstream of the nozzle. For the inlet condition the pulsating mass flow rate was used.

The outlet boundary was placed in the far field (Figure 5.4). The size of this outflow region is $50 \cdot d_{egp}$ in axial direction and $40 \cdot d_{egp}$ in radial direction. In this region the cells are stretched near to the outlet surface in order to alleviate reflections. The non-reflecting zone of Freund was recently modified and tested. For this case its usage is limited somewhat but still advantageous. At the outlet surface at $x=5$ $m$ the static pressure outlet condition is used and the surface is inclined based on the observation explained next. In order to obtain a statistically steady solution before applying the excitation at the inlet a long calculation on the multigrid level 4 (coarsest mesh) and 3 was carried out. The entropy waves generated by the transient of the initialization must be advected through the burner plenum and the combustion chamber and finally out of the system. This needed a relative long time as the velocity behind the baffle plates is quite small. After the acoustic waves generated also by the transient of the initialization were decayed, it was detected, that acoustic waves of a discrete frequency were amplified to extreme high amplitudes. The wave length coincided width the length of the computational domain. After the outlet surface was slanted these standing waves decayed.

At the surfaces the no-slip boundary condition and an adiabatic wall are imposed. For the first grid point normal to the wall $y^+<1$ is obtained.

Similar to the case of the single chamber no symmetry or periodic condition could be used.

For the distribution of the control volumes a very important aspect was to apply the findings of the investigations of the single resonator. Thus much more computational cells were arranged in the regions of the resonator neck and of the exhaust gas pipe, respectively, and in this case around the baffle plates. The final version of the mesh consists of approx. $27 \cdot 10^6$ control volumes distributed among 612 blocks.

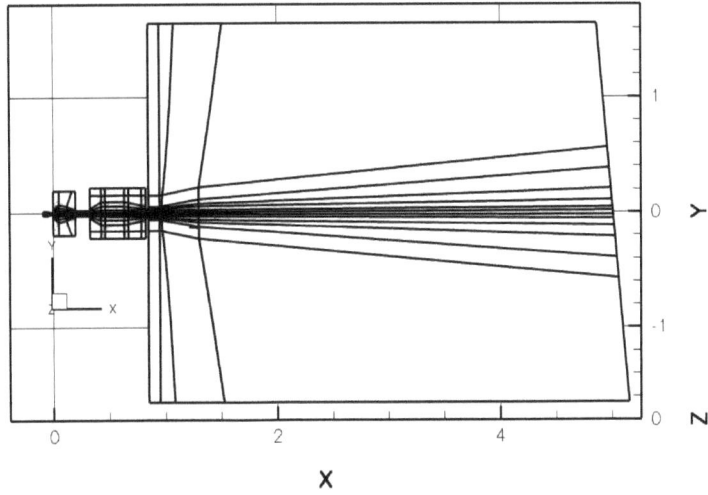

**Figure 5.4:** The computational domain with block structure in the symmetry plane

The spatial discretization is a second-order accurate central difference formulation. The temporal integration is carried out with a second-order accurate implicit dual-time stepping scheme. For the inner iterations the 5-stage Runge-Kutta scheme was used. The time step was $\Delta t = 2 \cdot 10^{-6}$ s. The Smagorinsky-Lilly, the dynamic Smagorinsky and the MILES approach were tested to modelling the SGS structures. The results show no significant differences in the predicted damping parameters like in the case of the single resonator (see Section 4). By means of the full multigrid method it was possible to study the effect of grid refinement.

## 5.2 Flow Properties in the Coupled System

The Reynolds number based on the mean bulk velocity is about $8.1 \cdot 10^4$ in the nozzle and in the resonator necks. This value is more than the double of the Reynolds number in the case of the single resonator. The visualization of the flow by means of the vorticity magnitude is depicted in Figure 5.5 and the $Q$-criterion in Figure 5.6. The plots demonstrate the high shear in the resonator neck and in the exhaust gas pipe, which agree well with the results of the single resonator. The high shear is due to the high acceleration of the pulsation. The deceleration phase of the pulsation period induces a strong separation of vortices from the shear layer into the middle part of the pipes and makes the flow highly turbulent there. This increases the momentum transport crosswise to the mean flow and implies an increasing of the hydraulic resistance.

The impingement of the jets on the baffle plates generates additional shear layers and vortex shedding which increase the damping effect. Behind the plates the flow motion is very slow and the flow is strongly accelerated again at the inflow into the resonator neck and into the exhaust gas pipe, respectively. The three-dimensional plot demonstrates in addition the influence of the observation windows. A part of the flow diverted by the baffle plate impinges on the lower part of the window and is diverted in negative $x$ direction.

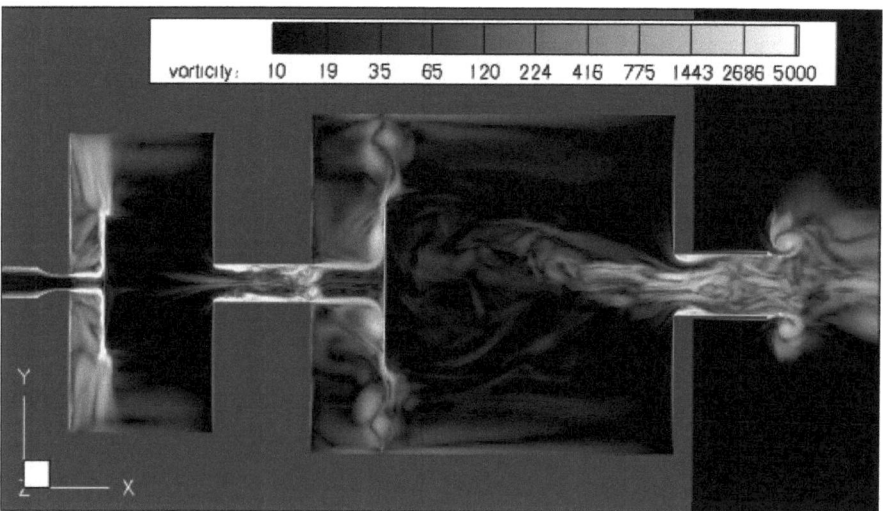

**Figure 5.5:** Distribution of vorticity in the symmetry plane

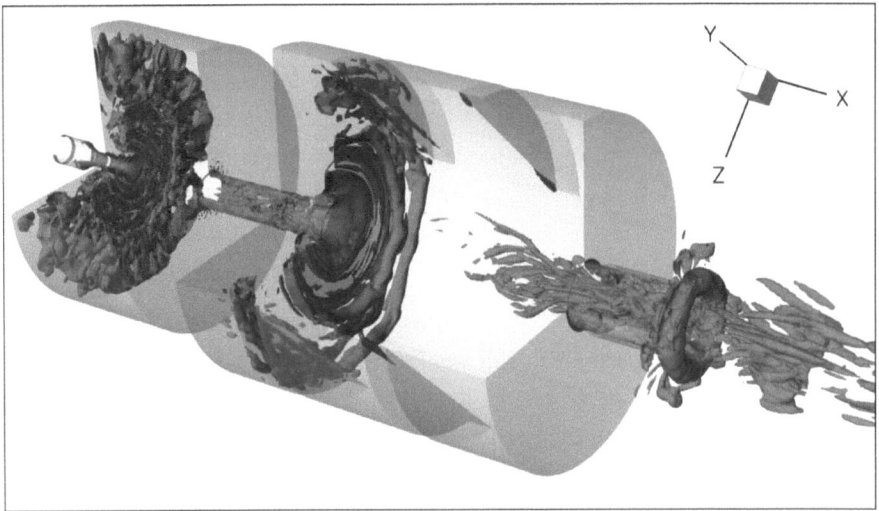

**Figure 5.6:** Iso-surfaces of the $Q$-criterion at the level of $10^4\ s^{-2}$

## 5.3 Comparison of the Results with Experimental Data

At the start of the numerical investigations the experiments were in progress and no data for comparison were available. As the resonant characteristics of the given configuration were unknown the eigenfrequencies of the components had been calculated first with the undamped Helmholtz resonator model as rough estimation. On the basis of this data 10 LESs were carried out on the coarsest mesh level. These calculations were relative fast because of the moderate number of grid points. Fortunately, the estimations were acceptable and the computations could be continued on the finer grid levels.

The results of the computation on different grid levels are plotted in Figure 5.7 and Figure 5.8. The difference in the resonance characteristics on the finest and second finest grid is negligible. It was tested only at the lower resonant frequency, at the highest amplitude ratio, because the calculation on the finest mesh is very time consuming. The higher is the amplitude ratio the higher are the demands on the mesh. This result shows that the flow phenomena, which are influencing the damping, are adequate resolved on the second finest mesh. It is important to take into consideration that the mesh was optimized on the results of the investigation of the single resonator.

The plotted results in Figure 5.7 and Figure 5.8 show generally a very good prediction of the resonance frequencies and of the phase shift, respectively. In Figure 5.7 there is, however, a discrepancy of approx. 20% in the prediction of the amplitude ratio at the highest peak, at $f_{ex}$=28 $Hz$. As discussed above, there is a significant shearing on the baffle plates. Unfortunately, in the experiments the plates were perforated to achieve the best velocity distribution at the outlet. In the simulations the wall condition was used for the plates. The resolution of the holes would yield a tremendous large number of grid points. A boundary condition which can simulate this effect was not available.

For steady flows the perforated wall can be simulated by means of the resistance coefficient expressed by the pressure loss and the reference dynamic pressure [49]. This approach was implemented and tested successfully. However, the interaction of the pulsating flow and the perforated plates cannot be covered by this relation. The description of this phenomenon can be handled only by dynamic equations. The pulsating flow in the holes is very similar to the one in the resonator neck between the burner plenum and the combustion chamber in the coupled resonator. However, in this case the flow in the holes remains laminar as the maximal Reynolds number was estimated to about 300. The drag coefficient can be described by means of the $1^{sth}$ or $2^{nd}$ problem of Stokes, if the flow direction is changed, respectively [122]. The modelling of the unsteady flow around a perforated plate could be therefore a considerably part of a PhD thesis and cannot be handled within the scope of this work.

As the geometry data were received, it was not possible to replace the plates any more. Probably this difference plays the major role in the under-prediction of the amplitude ratio.

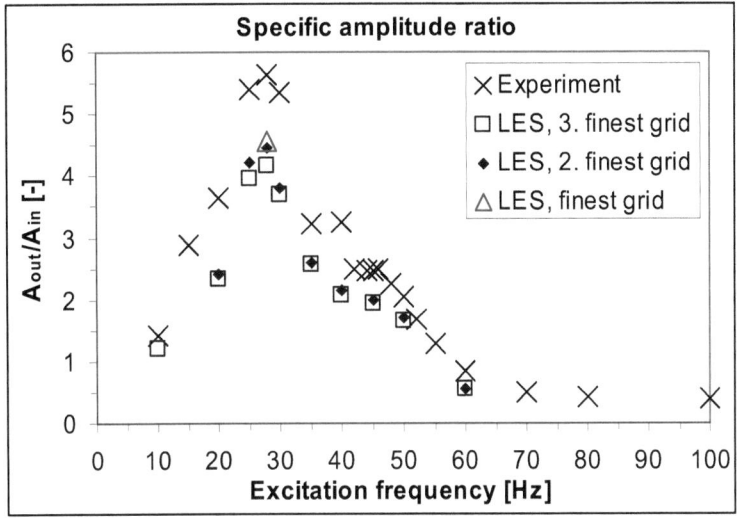

**Figure 5.7:** Amplitude response of the coupled resonators

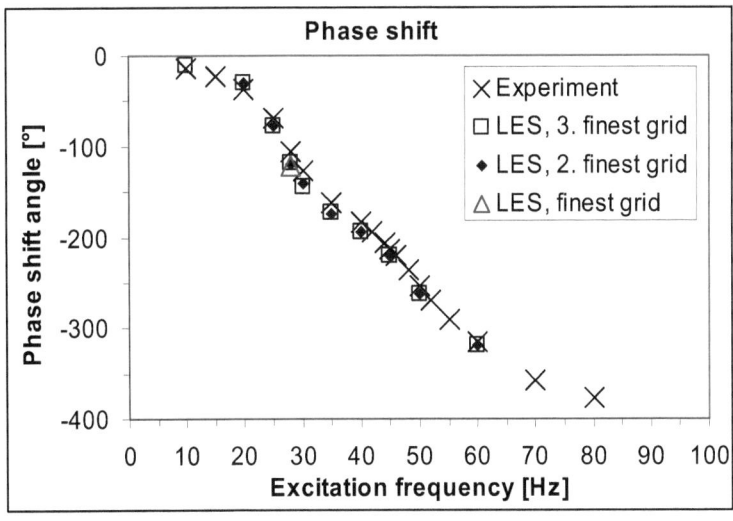

**Figure 5.8:** Phase transfer function of the coupled resonators

## 6 On the Source of Combustion Instabilities

As preparation for the simulations of the single resonator with roughness a calculation with constant mass flow rate at the inlet was carried out. The purpose of this simulation was to get a fully developed flow in all regions of the combustion chamber. The calculations with excitation at the inlet could use then the solution as initialization. In this case the computation could be carried out in a long time by means of computing on the new supercomputer HP-XC4000.

The experiences of the earlier investigations showed that the transient waves generated at the start of the computation should be decayed before starting the excitation in order to get the real system response at the outlet. The calculations are initialized in general with homogeneous distribution of each variable. This produces quite strong transient waves at the start. In SPARC the full multigrid method is implemented, which implies also grid sequencing. This method allows for getting a statistically steady state solution much faster. The waves generated by the initialization decay relative fast on the coarsest grid level. Additionally the computation is very fast because of the significantly reduced number of control volumes. For the combustion chamber four grid levels were used. Based on the earlier experiences, the excitation was started from the case with constant mass flow rate only on the third grid level, i.e. on the second finest mesh.

The mass flow rate signal at the outlet of the exhaust gas pipe was used to monitoring the decaying of the transient waves. As soon as an almost constant mass flow rate was reached the computation could be continued on the second coarsest grid level. The extrapolation of the solution from the coarser on the next finer grid level produces also transient waves because of the sudden change of the shear stress at the walls. These waves are much smaller than the waves generated at the initialization but they are still considerable on the second coarsest grid level.

The mass flow rate signal at the outlet showed the decaying of the transient waves on the second coarsest grid level but later a certain amount of pulsation was observed and it decayed not at all (Figure 6.1). The amplitude of this pulsation was not negligible and the frequency of the dominating wave was approximately at the eigenfrequency of the combustion chamber. Therefore the computation without excitation was continued on the finer grid levels also.

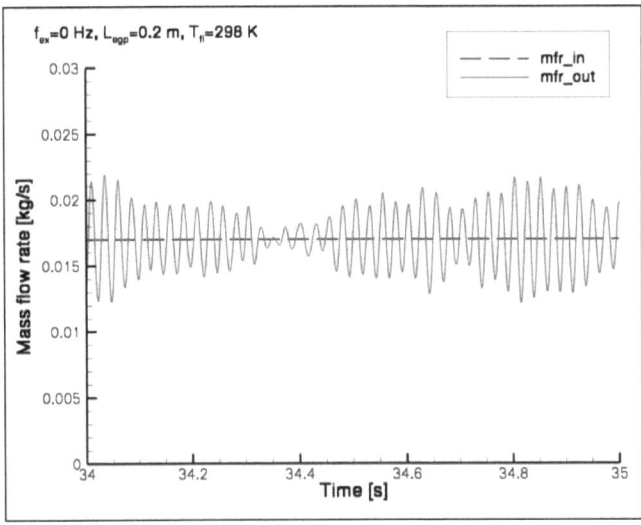

**Figure 6.1:** Mass flow rate of the single resonator without excitation at the inlet

The frequency spectrum plotted in Figure 6.2 is computed from the mass flow rate signal on the second finest and finest mesh. As the mass flow rate was computed through integration over the whole cross section of the exhaust gas pipe the turbulent fluctuations were mostly filtered out.

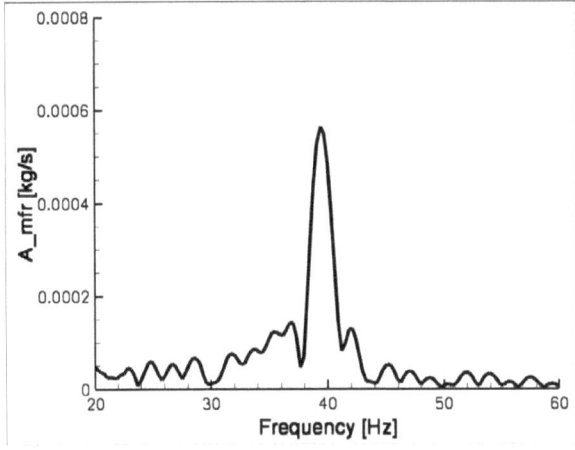

**Figure 6.2:** Frequency spectrum of the outlet mass flow rate of the single resonator

For this computation the time step was set to $\Delta t=10^{-4}$ s, a relative high value to get a better resolution of the spectra in the low frequency range. The samples were taken in each time step, thus the sampling frequency was 10 *kHz* furthermore the sampling length was 32768. The peak at 39 *Hz* agrees very good with the response function of the combustion chamber in Figure 4.18. Recent investigation showed that the resonance frequency can be captured already on the coarsest grid level and the signal of the solution on the second coarsest grid can already predict the resonance frequency quite accurately.

In [14] it was shown that the mass flow rate signal at the outlet and the pressure signal in the combustion chamber can be used as output signal equivalently i.e. the pulsation of the mass flow rate indicates a pulsation of the pressure in the chamber. The Fourier transform of the pressure signal measured at the middle of the side wall of the chamber gives the same result.

The mass flow rate at the inlet for this calculation was kept on a constant value. There was no external excitation in this computation and no turbulence at the inlet was described. The only possible forcing of the pulsation could arise from the turbulent motions inside the combustion chamber. The inflow into the chamber is a jet with strong shear layer which generates a broad band spectrum of turbulent fluctuations (Figure 4.12). The combustion chamber then amplifies the pressure fluctuations generated by turbulence at its eigenfrequency.

In order to investigate the effect of periodic flow instabilities further calculation with $\bar{m}=0.034$ *kg/s* and $\bar{m}=0.0136$ *kg/s* were carried out. The spectra of the mass flow rate of these calculations gave the same distribution in the low frequency range except the amplitude of the pulsation was changing proportional to the mean mass flow rate.

Based on these results the mass flow rate in the coupled resonators was also investigated. In Figure 6.3 the frequency spectrum of the mass flow rate signal on the second finest mesh is exhibited. The peaks at 27 *Hz* and 54 *Hz* correspond with the eigenfrequencies of the coupled system, which can be read e.g. from Figure 5.8 at the phase shift angle 90° and 270°, respectively.

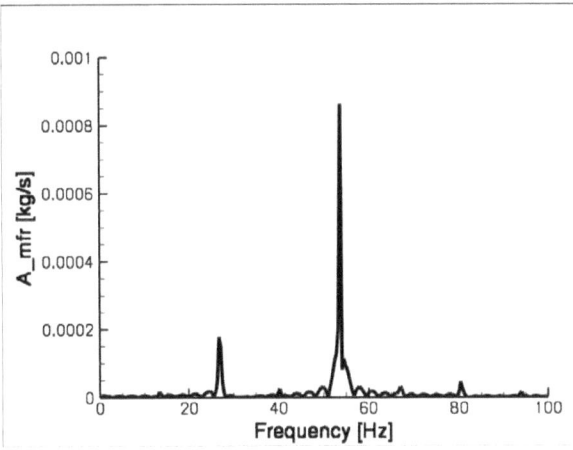

**Figure 6.3:** Frequency spectrum of the outlet mass flow rate of the coupled resonators

There are some possible mechanisms listed in the literature, which could trigger self-excited instabilities in combustion systems, but they are not sufficiently understood [14], [58], [104], [111]. An important achievement of this simulation is that the pressure in the combustion chamber can pulsate already without any external excitation e.g. compressor or other incoming disturbances from ambient or even periodic flow instabilities depending on the design of the burner. Thus the flame is also pulsating. The amplitude of this pulsation will be amplified to the limit cycle if the time lag of the flame changed so that the pressure fluctuation and the heat release fluctuation meet the Rayleigh criterion.

The results of the earlier investigations show that the pulsation and the high shear in the resonator neck produce highly anisotropic swirled flow. Therefore it is unlikely that a URANS simulation can render such flow reliably. Furthermore if the turbulence is modelled statistically, it cannot excite the flow in the combustion chamber. The use of LES for the investigation of combustion instabilities is essential.

For the analytical model the eigenfrequency of the system is an important input parameter. If the geometry is rather simple the undamped Helmholtz resonator model can be used. Further important achievement of the present computation is that the eigenfrequency of the system with geometry of high complexity can be predicted without an additional modal analysis. The calculation with constant mass flow rate is a preparation for the investigation with excitation at the inlet. Latter calculation provides the damping ratio for the analytical model.

## 7 Computational Effort

The calculations have been performed using the IBM RS/6000 SP of the Scientific Supercomputing Centre (SSC) Karlsruhe, the Hitachi SR8000-F1 of LRZ Munich and the same type of the HLRS of the University of Stuttgart, the CRAY Opteron-cluster of the HLRS, the HP XC6000 and HP XC4000 of the SSC Karlsruhe and the Opteron-cluster of the Department of Fluid Machinery at the University of Karlsruhe.

The computation of one cycle of the combustion chamber took approx. one week on the Opteron-cluster of the Department on 22 processors. On the HP XC6000 a typical calculation of the combustion chamber was running for about 300 hours (elapsed time) using 32 processors.

The coupled system was computed on the HP XC4000. The computation of one cycle on the second finest grid level took approx. 3 days and on the finest grid level approx. one month using 108 processors. The parallel efficiency on this system was about 98%.

# 8 Conclusions and Outlook

The lean premixed combustion allows for reducing the production of thermal $NO_x$, therefore it is largely used in stationary gas turbines and for other industrial combustion. Lean premixed combustors are, however, prone to combustion instabilities with both low and high frequencies. For the prediction of the stability of technical combustion systems the knowledge of the periodic-non-stationary mixing and reacting behaviour of the applied flame type and a quantitative description of the resonance characteristics of the gas volumes in the combustion chamber is conclusively needed.

In the frame of this work the numerical investigation of the non-reacting flow in a Helmholtz resonator-type model combustion chamber and in a coupled system of burner and combustion chamber is presented. The work was a part of series of investigations to determine the stability limits of combustion systems. The resonance characteristics of the combustion systems were calculated using Large Eddy Simulation. The results are in good agreement with the experimental data and a reduced physical model, which was developed by Büchner to describe the resonant behaviour of a damped Helmholtz resonator-type combustion chamber.

The solution of the fully compressible Navier-Stokes equations was essential to capture the physical response of the pulsation amplification, which is mainly the compressibility of the gas volume in the chamber. Viscous effects play a crucial role in the oscillating boundary layer in the neck of the Helmholtz resonator and, hereby, in the damping of the pulsation. The pulsation and the high shear in the resonator neck produce highly anisotropic turbulent flow. Therefore it is improbable that an URANS simulation can render such flow reliably.

The flow in the single resonator was analyzed in details. The simulations confirmed the basic assumptions made by the analytical model. The pulsating flow in the exhaust gas pipe was compared with experiments from the literature and good agreement was observed. It was found that the logarithmic wall function is not applicable for pulsating flows. The main damping mechanisms of the pulsation were investigated and proved that the oscillating boundary layer plays the major role. The comparison of the LES results with the measurements sheds light on the role of the wall roughness in the resonator neck, which affects significantly the damping in the system. The influence of the surface quality of the wall was not investigated systematically before. The result of the simulation with SEM shows that the inlet turbulence has no effect on the system response in the case of the investigated configuration.

The investigations showed that it is necessary to consider not only the single components at a technically relevant combustion system, but the behaviour of the coupled system, which can be completely different from the resonance characteristics of the single components.

The numerical investigations showed the capability of predicting the damping factor accurately by means of LES for a single resonator and for coupled systems as well. Using the full multigrid method it was possible to study the effect of the grid refinement. The computational domain of the coupled system is significantly larger than the domain of the single resonator. The results show, however, that the resonant characteristics are predicted sufficiently good already on the second finest grid level of the optimized mesh of the coupled system. The number of the computational cells on the second finest grid level is about the same as on the finest mesh of the single combustion chamber. Therefore it is expected that the prediction of the damping parameter of a more complex combustion system by LES is realizable within the limits of the computational resources. The results can, however, strongly depend on special boundary conditions as roughness or perforated plates. The effect of roughness is simulated with the Discrete Element Method successfully.

The investigation of the case without excitation showed that the frequency spectrum of the mass flow rate signal at the outlet provide a peak at the eigenfrequency of the combustion chamber. The only possible forcing of this pulsation was the turbulent fluctuations generated by the jet in the combustion chamber. The broadband excitation of the turbulent flow is then can be amplified by the flame and can produce a broadband background heat release rate oscillation as detected also in [149]. If the turbulence is modelled statistically (URANS), it cannot excite the flow in the combustion chamber. The use of LES for the investigation of combustion instabilities is essential.

It is expected that the experiences and models obtained in the Subproject A7 can be used in the future project of the CRC 606, where it is planed to extent the methodology for the modular combustion system with multiple burner units.

Based on the results of the single resonator, which show the important effect of the surface roughness of the wall in the exhaust gas pipe, it is proposed for Helmholtz resonators used for damping of oscillations, that wall roughness should be increased in the resonator neck e.g. by means of a perforated tube inserted in it. Furthermore based on the investigations of the coupled system it is proposed that as damping device coupled resonators can be used more efficiently than a single Helmholtz resonator. If they are tuned correctly they can provide a much wider range of high attenuation.

In the case of the single resonator it was shown that the pulsating exhaust gas jet provides a much higher entrainment. Pulsating jets are used for better heat transfer [42], [138] or thrust augmentation [99] but they can have also detrimental effect [46]. The application of pulsating jets in pneumatic or hydraulic conveying systems was not found in the literature. It is proposed here to investigate pulsating jets in jet pumps, especially of gas jet pumps, in which the pulsation can be amplified by a resonator (which is the combustion chamber in the present case).

# References

[1] Abouali, M., Geurts, BJ., Gieske, A.: Atmospheric Boundary Layers Over Rough Terrain. *ERCOFTAC*, **72**, pp. 7-11, 2007.

[2] Annaswamy, A.M., Ghoniem, A.F.: Active Control of Combustion Instability: Theory and Practice. *IEEE Control Systems Magazine*, **22** (6), pp. 37-54, 2002.

[3] Apsley, D.: CFD Calculation of Turbulent Flow with Arbitrary Wall Roughness. *Flow Turbulence and Combustion*, **78**, pp. 153-175, 2007.

[4] Arnold, G., Büchner, H.: Modelling of the Transfer Function of a Helmholtz-Resonator-Type combustion chamber. In *Proceedings of the European Combustion Meeting 2003* (ECM2003), Orléans, France, 2003.

[5] Aupoix, B., Spalart, P.R.: Extensions of the Spalart-Allmaras Turbulence Model to Account for Wall Roughness. *International Journal of Heat and Fluid Flow*, **24**, pp. 454–462, 2003.

[6] Baade, P.: Selbsterregte Schwingungen in Gasbrennern. *Klima- + Kälteingenieur*, Nr. 4/74, Teil 6, pp. 167-176, 1974.

[7] Baldwin, B., Lomax, H.: Thin Layer Approximation and Algebraic Model for Separated Turbulent Flows. *AIAA Paper* 78-257, 1978.

[8] Bellucci, V., Schuermans, B., Nowak, D., Flohr, P., Paschereit, C.O.: Thermoacoustic Modeling of a Gas Turbine Combustor Equipped with Acoustic Dampers. *Journal of Turbomachinery*, **127** (2), pp. 372-379, 2005.

[9] Bockhorn, H., Fröhlich, J., Suntz, R.: SFB606 – a German Research Initiative on Unsteady Combustion. *ERCOFTAC Bulletin*, **59**, pp. 40-44, 2003

[10] Boris, J. P., Grinstein, F. E., Oran, E. S., Kolbe, R. L.: New Insights into Large-Eddy Simulation. *Fluid Dynamics Research*, **10**, pp. 199-228, 1992.

[11] Boussinesq, J.V.: Théorie de l'Écoulement Tourbillant. *Mem. Présentés par Divers Savants Acad. Sci. Inst. Fr.*, **23**, pp. 46-50, 1877.

[12] Bradshaw, P.: Compressible Turbulent Shear Layers. *Annual Review of Fluid Mechanics*, **9**, pp. 33-54, 1977.

[13] Brandt, A.: Multi-Level Adaptive Computations in Fluid Dynamics. *AIAA Paper* 79-1455, 1979.

References

[14] Büchner, H.: *Strömungs- und Verbrennungsinstabilitäten in technischen Verbrennungssystemen*. Professorial dissertation, University of Karlsruhe, 2001.

[15] Büchner, H., Bockhorn, H., Hoffmann, S.: Aerodynamic Suppression of Combustion-Driven Pressure Oscillations in Technical Premixed Combustors. In *Proceedings of Symposium on Energy Engineering in the 21st Century* (SEE 2000), Hong Kong, China, Ping Cheng (editor), Begell House, New York, **4**, pp.1573-1580, 2000.

[16] Büchner, H., Külsheimer, C.: Untersuchungen zum frequenzabhängigen Mischungs- und Reaktionsverhalten pulsierender, vorgemischter Drallflammen. *GASWÄRME International*, **46** (2), pp. 122-129, 1997.

[17] Bühler, S.: *Numerische Untersuchung von Wandrauhigkeitsmodellen*. Master thesis, Department of Fluid Machinery, University of Karlsruhe, 2007.

[18] Cebecci, T., Chang, K.C.: Calculation of Incompressible Rough-Wall Boundary-Layer Flows. *AIAA Journal*, **16**, pp. 730-735, 1978.

[19] Culick, F.E.C.: Nonlinear Growth and Limiting Amplitude of Acoustic Oscillations in Combustion Chambers. *Combustion Science and Technology*, **3** (1), pp. 1-16, 1971.

[20] Deuker, E.: *Ein Beitrag zur Vorausberechnung des akustischen Stabilitätsverhaltens von Gasturbinenbrennkammern mittels theoretischer und experimenteller Analyse von Brennkammerschwingungen*. PhD thesis, Universität of RWTH Aachen, 1994.

[21] Doerffer, P., Namiesnik, K., Magagnato, F.: Flow Simulation at Shock Wave Triple Point. *TASK-Quarterly*, **5** (4), pp. 549-556, 2001.

[22] Doerffer, P., Szulz, O., Magagnato, F.: Shock Wave Boundary Layer Interaction in Forced Shock Oscillations. *Journal of Thermal Science*, **12** (1), pp. 10-15, 2003.

[23] El-Askary, W.A., Ewert, R., Schröder, W.: On Large Eddy Simulation as a Tool to Predict Acoustical Fields. In T. Hüttl, C. Wagner, J.W. Delfs edit., Proceedings of International Workshop on *LES for Aeroacoustics*, DGLR-Report 2002-03, 7-8 October 2002, German Aerospace Center, DLR, Göttingen, Germany, 2002.

# References

[24] Fang, Y., Menon, S.: A Two-Equation Subgrid Model for Large-Eddy Simulation of High Reynolds Number Flows. *AIAA*-06-0116, 2006.

[25] Ferziger, J.H., Perić, M.: *Computational Methods for Fluid Dynamics*. Springer, Berlin, 3$^{rd}$ Edition, ISBN 3-540-42074-6, 2002.

[26] Franke, J.: *Untersuchungen zur Grobstruktursimulation kompressibler Strömungen mit der Volumenfiltermethode auf bewegten Gittern*. PhD thesis, University of Karlsruhe, 1998.

[27] Freund, J.B.: Proposed Inflow/Outflow Boundary Condition for Direct Computation of Aerodynamic Sound. *AIAA-Paper* **35** (4), pp. 740-742, 1997.

[28] Fritz, W., Magagnato, F., Rieger, H.: Computational Methods for Prediction of Turbulent Flows around Automobiles. In *Proceedings of the 3$^{rd}$ International Conference of Innovation and Reliability in Automotive Design and Testing*, 1992.

[29] Fröhlich, J., von Terzi, D.: Hybrid LES/RANS Methods for the Simulation of Turbulent Flows. *Progress in Aerospace Sciences*, **44**, pp. 349-377, 2008.

[30] Germano, M., Piomelli, U., Moin, P., Cabot, W.H.: A Dynamic Subgrid-Scale Eddy Viscosity Model. *Physics of Fluid*, **3** (7), pp. 1760-1765, 1991.

[31] Giauque, A., Selle, L., Gicquel, L., Poinsot, T., Büchner, H., Kaufmann, P., Krebs, W.: System Identification of a Large-Scale Swirled Partially Premixed Combustor Using LES and Measurements. *Journal of Turbulence*, **6**, N 21, 2005.

[32] Geurts, B.J.: *Elements of Direct and Large-Eddy Simulation*. R.T. Edwards, Philadelphia, USA, 2004.

[33] Girimaji, S.S.: Partially-Averaged Navier-Stokes Model for Turbulence: A Reynolds-Averaged Navier-Stokes to Direct Numerical Simulation Bridging Method. *Journal of Applied Mechanics*, **73** (3), pp. 413-421, 2006.

[34] Gropp, W., Lusk, E., Thakur, R.: *Using MPI-2: Advanced Features of the Message-Passing Interface*. MIT-Press, Cambridge, ISBN 0-262-057133-1, 1999.

[35] Guyot, D., Taticchi Mandolini Borgia, P., Paschereit, C.O.: Active Control of Combustion Instability Using a Fluidic Actuator. In *Proceedings of the 46$^{th}$ AIAA Aerospace Sciences Meeting and Exhibit*, 7-10 January 2008, Reno, Nevada, AIAA 2008-1058.

[36] Gysling, D.L., Copeland, G..S., McCormick, D.C., Proscia, W.M.: Combustion System Damping Augmentation with Helmholtz Resonators. *ASME Journal of Engineering for Gas Turbines and Power*, **122** (2), 269-275, 2000.

[37] Hantschk, C.-C.: *Numerische Simulation selbsterregter thermoakustischer Schwingungen*. Fortschritt-Berichte VDI, Reihe 6, Nr. 441, 2000

[38] Heidmann, M. F., Povinelli, L. A.: An Experiment on Particulate Damping in a Two-Dimensional Hydrogen-Oxygen Combustor. *NASA TM* X-52359, 1967

[39] Higgins, B.: On the Sound Produced by a Current of Hydrogen Gas Passing Through a Tube. *Journal of Natural Philosophy, Chemistry and the Arts*, **1**, pp. 129-131, 1802.

[40] Hinze, J.O.: *Turbulence*. McGraw-Hill Book Company, New York, 1959.

[41] Hirsch, C.: *Numerical Computation of Internal and External Flows*, Volume 1 and 2. John Wiley & Sons, Chichester, 1988.

[42] Hofman, H.M., Movileanu, D.L., Kind, M., Martin, H.: Influence of a Pulsation on Heat Transfer and Flow Structure in Submerged Impinging Jets. *International Journal of Heat and Mass Transfer*, **50** (17-18), pp. 3638-3648, 2007.

[43] Hu, F.Q.: Absorbing Boundary Conditions. *International Journal of Computational Fluid dynamics*, **18** (6), pp. 513-522, 2004.

[44] Huang, P.G., Coleman, G.N., Bradshaw, P.: Compressible Turbulent Channel Flows: DNS Results and Modelling. *Journal of Fluid Mechanics*, **305**, pp. 185-218, 1995.

[45] Hudson, G.E.: Thrust on a Piston Driven Half-Open Tube. *The Journal of the Acoustical Society of America*, **27** (3), pp. 406-416, 1955.

[46] Hulet, C., Briens, C., Berruti, F., Chan, E.W., Ariyapadi, S.: Entrainment and Stability of a Horizontal Gas-Liquid Jet in a Fluidized Bed. *International Journal of Chemical Reactor Engineering*, 1, Article A60, 2003.

[47] Hummel, F., Lötzerich, M.: Surface Roughness Effects on Turbine Blade Aerodynamics. *Journal of Turbomachinery*, **127**, pp. 453-461, 2005.

# References

[48] Hunt, J.C.R., Wray, A.A, Moin, P.: Eddies, Streams and Convergence Zones in Turbulent Flows. *Report CTR-S88*, Center for Turbulence Research, 1988.

[49] Idelchik, I.E.: *Handbook of Hydraulic Resistance*, 2$^{nd}$ Edition. Springer, 1986.

[50] International Energy Agency, *Key World Energy Statistics 2008*, www.iea.org

[51] Iwamoto, K.: Database of Fully Developed Channel Flow. *THTLAB Internal Report*, No. ILR-0201., THTLAB, Dept. of Mech. Eng., The Univ. of Tokyo, 2002.

[52] Iwamoto, K., Suzuki, Y., Kasagi, N.: Reynolds Number Effect on Wall Turbulence: Toward Effective Feedback Control. *Int. J. Heat and Fluid Flow*, **23**, pp. 678-689, 2002.

[53] Jameson, A.: Transonic Flow Calculations. *MAE Report* 1651, Mechanical and Aerospace Engineering Dept., Princeton Univ., Princeton, NJ, 1983.

[54] Jameson, A., Schmidt, W., Turkel, E.: Numerical Solution for the Euler Equations by Finite Volume Methods Using Runge-Kutta Time Stepping Schemes. *AIAA Paper* 81-1259, 1981.

[55] Jarrin, N., Benhamadouche, S., Laurence, D., Prosser, R.: A Synthetic-Eddy-Method for Generating Inflow Conditions for Large-Eddy Simulations. In *Proceedings of the Symposium on Hybrid RANS-LES Methods*, Stockholm, 14-15 July, 2005.

[56] Jensen, B. L., Sumer, B. M., Fredsoe, J: Turbulent Oscillatory Boundary Layers at High Reynolds Numbers. *Journal of Fluid Mechanics*, **206**, pp.265-297, 1989.

[57] Johnson, C.E., Neumeier, Y., Lieuwen, T., Zinn, B.T.: Experimental Determination of the Stability Margin of a Combustor Using Exhaust Flow and Fuel Injection Rate Modulations. In *Proceedings of the Combustion Institute*, **28**, pp. 757-764, 2000.

[58] Joos, F.: *Technische Verbrennung*. Springer, Berlin. ISBN-13 978-3-540-34333-2. (2006)

[59] Kaltenbach, H. J., Fatica, M., Mittal, R., Lund, T. S., Moin, P.: Study of Flow in a Planar Asymmetric Diffuser Using Large-Eddy Simulation. *Journal of Fluid Mechanics*, **390**, pp. 151-185, 1999.

[60] Kang, H. S., Chester, S., Meneveau, C.: Decaying Turbulence in an Active-Grid-Generated Flow and Comparison with Large-Eddy Simulation. *Journal of Fluid Mechanics*, **480**, pp. 129-160, 2003.

[61] Kim, K., Jones, C.M., Lee, J.G., Santavicca, D.A.: Active Control of Combustion Instabilities in Lean Premixed Combustors. In *Proceedings of the 6$^{th}$ Symposium on Smart Control of Turbulence*, Tokyo, 2005.

[62] Klein, M., Sadiki, A., Janicka, J.: A Digital Filter Based Generation of Inflow Data for Spatially Developing Direct Numerical or Large Eddy Simulation. *Journal of Computational Physics*, **186**, pp. 652-665, 2003.

[63] Külsheimer, C., Büchner, H., Leuckel, W., Bockhorn, H., Hoffmann, S.: Untersuchung der Entstehungsmechanismen für das Auftreten periodischer Druck-/Flammenschwingungen in hochturbulenten Verbrennungssystemen. *VDI-Berichte* 1492, pp. 463, 1999.

[64] Lefebvre, A.H.: The Role of Fuel Preparation in Low-Emission Combustion. *ASME Journal of Engineering for Gas Turbines and Power*, **117**, pp. 617-665, (1995).

[65] Lehmann, K.O.: Über die Theorie der Netztöne (thermisch erregte Schallschwingungen). *Annalen der Physik*, **421** (6), pp. 527-555, 1937.

[66] Lenormand, E., Sagaut, P., Phuoc, L.T., Comte, P.: Subgrid-Scale Models for Large-Eddy Simulations of Compressible Wall Bounded Flows. *AIAA Journal*, **38**, pp. 1340-1350, 2000.

[67] Lenz, W.: *Die dynamischen Eigenschaften von Flammen und ihr Einfluss auf die Entstehung selbsterregter Brennkammerschwingungen*. PhD thesis, University of Karlsruhe, 1980.

[68] Lieuwen, T.C.: Online Combustor Stability Margin Assessment Using Dynamic Pressure Data. *Journal of Engineering for Gas Turbines and Power*, **127** (3), pp. 478-482, 2005.

[69] Lieuwen, T.C., Yang, V.: *Combustion Instabilities in Gas Turbine Engines*. AIAA, ISBN-13 978-1-56347-669-3, 2005.

[70] Lilly, D. K.: The Representation of Small-Scale Turbulence in Numerical Simulation Experiments. In *Proceedings of the IBM Scientific Computing Symposium on Environmental Sciences*, Yorktown Heights, N.Y., IBM form no. 320-1951, White Plains, New York, pp.195-210, 1967.

[71] Lodahl, C.R., Sumer, B.M., Frese, J.: Turbulent Combined Oscillatory Flow and Current in a Pipe. *Journal of Fluid Mechanics*, **373**, pp. 313-348, 1998.

[72] Lohrmann, M., Arnold, G., Büchner, H.: Modelling of the Resonance Characteristics of a Helmholtz-Resonator-Type Combustion Chamber with Energy Dissipation, *Proceedings of the International Gas Research Conference (IGRC)*, Amsterdam, Netherlands, 2001.

[73] Lohrmann, M., Bender, C., Büchner, H., Zarzalis, N.: Scaling of Stability Limits by Use of Universal Flame Transfer Functions. In *Proceedings of the Joint Congress CFA/DAGA'04*, vol. 2, Strasbourg, France, 2004.

[74] Lohrmann, M., Büchner, H.: Prediction of Stability Limits for LP and LPP Gas Turbine Combustors. *Combustion Science and Technology*, **177** (12), pp. 2243-2273, 2005.

[75] Lohrmann, M., Büchner, H.: Scaling of Stability Limits of Lean-Premixed Gas Turbine Combustors. In *Proceedings of ASME Turbo Expo*, Wien, Austria, 2004.

[76] Lund, T., Wu, X., Squires, D.: Generation of Turbulent Inflow Data for Spatially-Developing Boundary Layer Simulations. *Journal of Computational Physics*, **140**, pp. 233-258, 1998.

[77] Magagnato, F.: Computation of Axisymmetric Base Flows with Different Turbulence Models. *AGARD Conference Proceedings*, 493, 1990.

[78] Magagnato, F.: KAPPA – Karlsruhe parallel program for aerodynamics. *TASK Quarterly*, **2**, pp. 215-270, 1998.

[79] Magagnato, F.: The Modelling of Unsteady Turbulent Flows in Turbomachines. *TASK Quarterly*, **5** (4), pp. 477-494, 2001.

[80] Magagnato, F.: Unsteady Flows Past a Turbine Blade Using Non-Linear Two-Equation Turbulence Model. In *Proceedings of the 3$^{rd}$ European Conference on Turbomachinery: Fluid Dynamics and Thermodynamics*, London, 1999.

[81] Magagnato, F.: *Untersuchung von linearen und nichtlinearen Wirbelviskositätsmodellen*. PhD thesis, Darmstadt, 1995.

[82] Magagnato, F., Bühler, S., Gabi, M.: Modelling the wall roughness for RANS and LES Using the Discrete Element Method. ICTAM 2008, Adelaide, Australia, 2008.

[83] Magagnato, F., Gabi, M.: A New Adaptive Turbulence Model for Unsteady Flow Field in Rotating Machinery. *International Journal of Rotating Machinery*, **8** (3), pp. 175-183, 2002.

[84] Magagnato, F., Gabi, M., Heidenreich, T., Velji A., Spicher, U.: Large Eddy Simulation (LES) with Moving Meshes on a Rapid Compression Machine: Part 2: Numerical Investigations Using Euler-Lagrange-Technique. *High-Performance Computing in Science and Engineering '07*, Nagel, W., Kröner and D. Resch. M. Eds. Springer Berlin Heidelberg, pp. 419-430, 2007.

[85] Magagnato, F., Pritz, B., Büchner, H., Gabi, M.: Prediction of the Resonance Characteristics of Combustion Chambers on the Basis of Large-Eddy Simulation. *International Journal of Thermal and Fluid Sciences*, **14** (2), pp. 156-161, 2005.

[86] Magagnato, F., Pritz, B., Gabi, M.: Calculation of a Turbine Blade at High Reynolds Numbers by Large-Eddy Simulation. *The 11$^{th}$ of International Symposium on Transport Phenomena and Dynamics of Rotating Machinery*, Honolulu, Hawaii, February 26 - March 2, 2006.

[87] Magagnato, F., Rachwalski, J., Gabi, M.: An Application of the Buffer Layer Technique to Computations of Flow in Turbomachinery. *Proceedings of the 12$^{th}$ International Conference on Fluid Flow Technologies*, Budapest, Hungary, 2003.

[88] Magagnato, F., Rachwalski, J., Gabi, M.: Numerical Investigation of the VKI Turbine Blade by Large Eddy Simulation. In *Proceedings of the 6$^{th}$ European Conference on Turbomachinery: Fluid Dynamics and Thermodynamics*, Lille, 2005.

[89] Magagnato, F., Sorgüven, E., Gabi, M.: Far Field Noise Prediction by Large Eddy Simulation and Ffowcs-Williams Hawkings Analogy. In *Proceedings of the 9$^{th}$ AIAA/CEAS Aeroacoustics Conference*, Hilton Head SC, AIAA Paper 2003-3206, 2003.

[90] Magagnato, F., Yong, Y., Gabi, M.: Large Eddy Simulation of a Spark-Ignition Engine Using a High-Pass Filtered Eddy-Viscosity Model. In *Proceedings of the 7$^{th}$ COMODIA International Conference on Modelling and Diagnostics*, Sapporo, Japan, 2008.

[91] Manna, M., Vacca, A.: Spectral Dynamic of Pulsating Turbulent Pipe Flow. *Computers and Fluids*, **37**, pp. 825-835, 2008.

# References

[92] Maruyama, T.: Optimization of Roughness Parameters for Staggered Arrayed Cubic Blocks Using Experimental Data. *Journal of Wind Engineering and Industrial Aerodynamics*, **46-47**, pp. 165–171, 1993.

[93] Menter, F. R.: Two-Equation Eddy Viscosity Turbulence Models for Engineering Applications. *AIAA Journal*, **32** (8), pp. 1598-1605, 1994.

[94] Miyake, Y., Tsujimoto, K., Agata, Y.: A DNS of a Turbulent flow in a Rough-Wall Channel Using Roughness Element Model. *JSME International Journal*, Series B, **43**, pp. 233-241, 2000.

[95] Moin, P., Wang, M.: Wall Modeling for Large-Eddy Simulation of Turbulent Boundary Layers. IUTAM Symposium on One Hundred Years of Boundary Layer Research, in *Proceedings of the IUTAM Symposium held at DLR-Göttingen*, Germany, August 12-14, 2004.

[96] Monfort, D., Benhamadouche, S., Sagaut, P.: Extended Wall-Function for Velocity and Temperature in Large-Eddy Simulation. *Turbulence, Heat and Mass Transfer 5, Proceeding of the International Symposium on Turbulence, Heat and Mass Transfer*, Dubrovnik, Croatia, September 25-29, 2006.

[97] Murray, R.M., Jacobson, C.A., Casas, R., Khibnik, A.I., Johnson, C.R., Jr., Bitmead, R., Peracchio, A.A., Proscia, W.M.: System Identification for Limit Cycling Systems: A Case Study for Combustion Instabilities. In *Proceedings of the American Control Conference*, Philadelphia, PA., 1998.

[98] Ni, R.-H.: Multiple-Grid Scheme for Solving the Euler Equations. *AIAA Journal*, **20** (11), pp. 1565-1571, 1982.

[99] Olcay, A.B., Krueger, P.S.: Measurement of Ambient Fluid Entrainment During Laminar Vortex Ring Formation. *Experiments in Fluids*, **44** (2), pp. 235-247, 2008.

[100] Panara, D., Porta, M., Dannecker, R., Noll, B.: Wall-Functions and Boundary Layer Response to Pulsating and Oscillating Turbulent Channel Flows. In *Proceedings of the $5^{th}$ International Symposium on Turbulence, Heat and Mass Transfer THMT06*, 2006.

[101] Pantle, I., Bárdossy, G., Gabi, M.: Numerischer Ansatz zur Simulation von Strukturbewegung induziert durch Fluidschwingungen. *34. Jahrestagung für Akustik*, DAGA, Ref. 387, 2008.

[102] Pantle, I., Gabi, M.: Discussion of CAA Calculation Methods for Internal Flows. In *Proceedings of the 2008 Congress and Exposition on Noise Control Engineering (INTERNOISE)*, Shanghai, China, 2008.

[103] Petsch, O., Pritz, B., Magagnato, F., Büchner, H.: Untersuchungen zum Resonanzverhalten einer Modellbrennkammer vom Helmholtz-Resonator-Typ. *Verbrennung und Feuerungen - 22. Deutscher Flammentag*, VDI-GET, VDI-Berichte Nr. 1888, pp. 507-512, 2005.

[104] Poinsot, T., Veynant, D.: *Theoretical and Numerical Combustion*. R.T. Edwards, Inc., Philadelphia, 2$^{nd}$ Edition. ISBN 1-930217-10-2, 2005.

[105] Priesmeier, U.: *Das dynamische Verhalten von Axialstrahl-Diffusionsflammen und dessen Bedeutung für selbsterregte Brennkammerschwingungen*. PhD thesis, Universität of Karlsruhe, 1987.

[106] Pritz, B., Magagnato, F., Gabi, M.: Inlet Condition for Large-Eddy Simulation Applied to a Combustion Chamber. In *Proceedings of the Conference on Modelling Fluid Flow (CMFF'06), The 13$^{th}$ International Conference on Fluid Flow Technologies*, Budapest, Hungary, September 6-9, 2006.

[107] Putnam, A.A.: *Combustion-Driven Oscillations in Industry*. American Elsevier, New York, 1971.

[108] Rachwalski, J., Magagnato, F., Gabi, M.: The Buffer Layer Technique Applied to Transonic Flow Calculations. *Symposium Transsonicum IV*, Göttingen, 2002.

[109] Rayleigh, J.W.S.: *The Theory of Sound*, Volume 2. Macmillan, London, 1878.

[110] Reynolds, O.: On the Dynamical Theory of Incompressible Viscous Fluids and the Determination of the Criterion. *Philosophical Transactions of the Royal Society of London, Series A*, **186**, pp. 123-164, 1895.

[111] Reynst, F.H.: *Pulsating Combustion*. Edited by M.W. Thring, Pergamon Press, Oxford, 1961.

[112] Richards, G.A., Thornton, J.D., Robey, E.H., Arellano, L.: Open-Loop Active Control of Combustion Dynamics on a Gas Turbine Engine. *Journal of Engineering for Gas Turbines and Power*, **129** (1), pp. 38-48, 2007.

[113] Richards, G.A., Robey, E.H.: Effect of Fuel System Impedance Mismatch on Combustion Dynamics. *Journal of Engineering for Gas Turbines and Power*, **130** (1), 011510, 2008.

[114] Richardson, E.G., Tyler, E.: The Transverse Velocity Gradient near the Mouth of Pipes in which an Alternating or Continuous Flow of Air is Established. *The Proceedings of the Physical Society*, **42** (231), pp.1-15, 1929.

[115] Rommel, D.: *Numerische Simulation des instationären, turbulenten und isothermen Strömungsfeldes in einer Modellbrennkammer.* Master thesis, Engler-Bunte-Institute, University of Karlsruhe, 1995.

[116] Russ, M., Büchner, H.: Berechnung des Schwingungsverhaltens gekoppelter Helmholtz-Resonatoren in technischen Verbrennungssystemen. *Verbrennung und Feuerung*, VDI-Berichte zum 23. Deutscher Flammentag, 2007.

[117] Sagaut, P.: *Large Eddy Simulation for Incompressible Flows.* Springer, Berlin, 2006.

[118] Sagaut P., Garnier E., Tromeur E., Larchevêque L., Labourasse E.: Turbulent Inflow Conditions for Large-Eddy Simulation of Compressible Wall-Bounded Flows. *AIAA Journal*, **42** (3), pp. 469-477, 2004.

[119] Sato, H., Nishidome, C., Kajiwara, I., Hayashi, A.K.: Design of Active Control System for Combustion Instability Using $H^2$ Algorithm. *International Journal of Vehicle Design*, **43**, pp. 322-340, 2007.

[120] Sattelmayer, T., Polifke, W.: A Novel Method for the Computation of the Linear Stability of Combustors. *Combustion Science and Technology*, **175** (3), pp. 477-497, 2003.

[121] Sattelmayer, T., Polifke, W.: Assessment of Methods of the Computation of the Linear Stability of Combustors. *Combustion Science and Technology*, **175** (3), pp. 453-476, 2003.

[122] Schlichting, H., Gersten, K.: *Grenzschicht-Theorie.* Springer, 9th edition, 1997.

[123] Schumann, U.: Subgrid-Scale Model for Finite Difference Simulation of Turbulent Flows in Plane Channels and Annuli. *Journal of Computational Physics*, **18**, pp. 376-404, 1975.

[124] Scotti, A., Piomelli, U.: Numerical Simulation of Pulsating Turbulent Channel Flow. *Physics of Fluid*, **13** (5), pp.1367-1384, 2001.

[125] Sexl, T.: Über den von E. G. Richardson entdeckten „Annulareffekt". *Zeitschrift für Physik*, **61**, pp. 349-362, 1930.

[126] *Shell Energy Scenarios to 2050*, Shell International BV, Amsterdam 2008 (www.shell.com/scenarios)

[127] Shih, T-H., Povinelli, L.A., Liu, N-S.: Application of Generalized Wall Function for Complex Turbulent Flows. *Journal of Turbulence*, **4**, 015, 2003.

[128] Smagorinsky, J.: General Circulation Experiments with the Primitive Equations. *Monthly Weather Review*, **91**, pp. 99-164, 1963.

[129] Spalart, P.R., Allmaras, S.R.: A One-Equation Turbulence Model for Aerodynamic Flows. *AIAA Paper* 92-0439, 1992.

[130] Spalart, P.R., Jou, W.H., Strelets, M., Allmaras, S.R.: Comments on the Feasibility of LES for Wings, and on a Hybrid RANS/LES Approach. In *Proceedings of the First AFOSR International Conference on DNS/LES*, Ruston, Louisiana, USA, 1997.

[131] Spalart, P.R., Shur, M.: On the Sensitization of Turbulence Models to Rotation and Curvature. *Aerospace Science and Technology*, **1** (5), 297-302, 1997.

[132] Speziale, C.G.: Computing Non-Equilibrium turbulent Flows with Time-Dependent RANS and VLES. *Fifteenth International Conference on Numerical Methods in Fluid Dynamics, Lecture Notes in Physics*, **490**, Springer, 1997.

[133] Stolz, S., Schlatter, P., Meyer, D., Kleiser, L.: High-Pass Filtered Eddy-Viscosity Models for LES. In *Direct and Large Eddy Simulation V*, Friedrich, R., Geurts, B.J., Métais, O. (eds.), Kluwer, Dordrecht, pp. 81-88, 2003.

[134] Swanson, R. C., Turkel, E.: Computational Fluid Dynamics: Multistage Central Difference Schemes for the Euler and Navies-Stokes Equations. *Von Karman Institute for Fluid Dynamics*, 1996.

[135] Tatsumi, S., Martinelli, L., Jameson, A.: Design, Implementation and Validation of Flux Limited Schemes for the Solution of the Compressible Navier-Stokes Equations. *AIAA Paper* 94-0647, 1994.

## References

[136]  Taylor, R.P., Coleman, H.W., Hodge, B.K.: Prediction of Turbulent Rough-Wall Skin Friction Using a Discrete Element Approach. *Journal of Fluids Engineering*, **107**, pp. 251-257, 1985.

[137]  *Tecplot® User's Manual*, Version 10. Tecplot, Inc., Bellevue, Washington, 2005.

[138]  Tesař, V., Trávníček Z.: Pulsating and Synthetic Impinging Jets. *Journal of Visualization*, **8** (3), pp. 201-208, 2005.

[139]  Tongue, B.H.: *Principles of Vibration*. Oxford University Press, Oxford, $2^{nd}$ edition, ISBN 0-19-514246-2, 2002.

[140]  Tran, N., Ducruix, S., Schuller, T.: Damping Combustion Instabilities with Perforates at the Premixer Inlet of a Swirled Burner. In *Proceedings of the Combustion Institute*, **32**, pp. 2917-2924, 2009.

[141]  Truckenbrodt, E.: *Fluidmechanik*. Springer, Berlin, 1980.

[142]  Tsuji, Y., Morikawa, Y.: Turbulent Boundary Layer with Pressure Gradient Alternating in Sign. *Aeronautical Quarterly*, **27**, pp.15-28, 1976.

[143]  Utyuzhnikov, S.V.: Generalized Wall Functions and Their Application for Simulation of Turbulent Flows. *International Journal for Numerical Methods in Fluids*, **47** (10-11), pp. 1323-1328, 2005.

[144]  Vincenti, W.G., Kruger, C.H.: *Introduction to Physical Gas Dynamics*. John Wiley and Sons, new York, 1965.

[145]  Von Neumann, J., Richtmyer, R.D.: A Method for the Numerical Calculation of Hydrodynamic Shocks. *Journal of Applied Physics*, **21**, p.232-237, 1950.

[146]  Wilcox, D.C.: *Turbulence Modeling for CFD*. DCW Industries, La Cañada, $2^{nd}$ edition, 1998.

[147]  www.ansys.com/products/icemcfd.asp

[148]  www.sfb606.uni-karlsruhe.de

[149]  Yi, T., Gutmark, E.J.: Online Prediction of the Onset of Combustion Instability Based on the Computation of Damping Ratios. *Journal of Sound and Vibration*, **310**, pp. 442-447, 2008.

[150]  Zierep, J.: *Grundzüge der Strömungslehre*. Springer, Karlsruhe, $6^{th}$ edition, 1997.

[151] Zierep, J., Bühler, K.: *Strömungsmechanik*. Springer, Berlin, ISBN 3-540-53827-5, 1991.

[152] Zinn, B.T.: Pulse Combustion Applications: Past, Present and Future. In *Unsteady Combustion*, Culick F. *et al.*, (eds.), NATO ASI Series, **306**, pp. 113-137, 1996.

[153] Zinn, B.T., Powell, E.A.: Nonlinear Combustion Instabilities in Liquid Propellant Rocket Engines. *Proceedings of the Combustion Institute*, **13**, pp. 491-502, 1970.

[154] Zou, Z., Xu, L.: Prediction of 3-D Unsteady Flow Using Implicit Dual Time Step Approach. *Acta Aeronautica et Astronautica Sinica*, **21** (1), pp 317-321, 2000.

## List of Tables

**Table 4.1:** Parameter set of the simulations .................................................................. 39

## List of Figures

**Figure 2.1:** Feedback loop of a combustion system ................................................... 7

**Figure 2.2:** The Helmholtz resonator and a mass-spring-damper system .................. 7

**Figure 3.1:** Mesh generation rules ............................................................................. 18

**Figure 3.2:** Distribution of the mean velocity of the turbulent boundary layer ........ 20

**Figure 3.3:** Distribution of the simulated cones ....................................................... 22

**Figure 3.4:** Flow over the flat plate. Drag coefficient versus $Re_x$ .............................. 23

**Figure 3.5:** Control of the mass flow rate based on the instantaneous mass flow rate difference ................................................................................................ 29

**Figure 3.6:** Control of the mass flow rate completed with a driving function ........... 29

**Figure 3.7:** The sketch of the computational domain ................................................ 30

**Figure 3.8:** The mean velocity profile in wall units (only each fourth data are plotted in the case of DNS and LES-FC) ................................................... 31

**Figure 3.9:** Temporal evolution of the averaged mass flow rate and forcing term: a) constant mass flow rate (LES-MC), b) constant force (LES-FC) ............................................................................................................ 32

**Figure 3.10:** Filtered RMS velocity profiles in wall coordinates (only each fourth DNS data are plotted) ................................................................... 32

**Figure 3.11:** Filtered Reynolds shear stress profile (only each fourth DNS data are plotted) ............................................................................................ 32

**Figure 3.12:** Evolution of the turbulence intensity of the spatially decaying isotropic turbulence ................................................................................. 36

**Figure 3.13:** Iso-surfaces of the $Q$-criterion at the level of 400 $s^{-2}$ ........................... 37

**Figure 4.1:** The sketch of the test rig and the analogy of the mass-spring-damper system and the combustion chamber as Helmholtz resonator .................. 38

**Figure 4.2:** Sketch of the computational domain and boundary conditions .............. 40

**Figure 4.3:** Third finest mesh extracted to the symmetry plane (distortions were caused by the extraction in Tecplot) ........................................................ 42

**Figure 4.4:** Instantaneous vortex structures and velocity vectors in the symmetry plane ........................................................................................................ 42

## List of Figures

**Figure 4.5:** One period of the mass flow rate and pressure signal of Case II ............ 43

**Figure 4.6:** Pressure at the symmetry axis and at the wall of the combustion chamber at different fractions of a period of pulsation (Case II) ............... 43

**Figure 4.7:** Velocity distributions at the half length of the exhaust gas pipe at different fractions of the period of pulsation ............................................. 45

**Figure 4.8:** Dissipation of the flow without excitation ............................................. 46

**Figure 4.9:** Over one period averaged dissipation of the flow with excitation near the resonant frequency .................................................................... 46

**Figure 4.10:** Distribution of the dissipation at the half length of the exhaust gas pipe in radial direction ............................................................................. 47

**Figure 4.11:** Dissipation in the region of the resonator neck; top: steady mean flow, bottom: pulsating flow ................................................................... 47

**Figure 4.12:** Iso-surfaces of the $Q$-criterion in the steady mean flow at the level of $2 \cdot 10^5 \ s^{-2}$ ........................................................................................... 48

**Figure 4.13:** Iso-surfaces of the $Q$-Criterion in the pulsating flow at the level of $2 \cdot 10^5 \ s^{-2}$ ........................................................................................... 49

**Figure 4.14:** Entrainment of the steady jet of Case I and the pulsating jet of Case II ....................................................................................................... 50

**Figure 4.15:** Entrainment of the pulsating jet of Case IV ........................................... 50

**Figure 4.16:** Distribution of the temperature at different fractions of the pulsation cycle (Case III at $f_{ex}$=54 $Hz$) ................................................. 51

**Figure 4.17:** Mass flow rate evolution to cycle limit .................................................. 52

**Figure 4.18:** Amplitude response of the combustion chamber .................................... 53

**Figure 4.19:** Phase transfer function of the combustion chamber ............................... 53

**Figure 4.20:** Distribution of the turbulent kinetic energy in the symmetry plane (Case I with SEM) .................................................................................. 54

**Figure 4.21:** Amplitude response of the combustion chamber .................................... 55

**Figure 4.22:** Phase transfer function of the combustion chamber ............................... 55

**Figure 4.23:** Iso-surfaces of the $Q$-criterion at the level of $2 \cdot 10^5 \ s^{-2}$ ......................... 55

**Figure 4.24:** Amplitude ratios from mass flow rates and pressure signal,

List of Figures

respectively .......................................................................................................... 56

**Figure 4.25:** Amplitude response of the combustion chamber ................................... 57

**Figure 4.26:** Phase transfer function of the combustion chamber ............................. 57

**Figure 4.27:** Velocity distributions at the half length of the exhaust gas pipe. a) Normalized U-profile of the non-pulsating and of the pulsating flow. b) U-profiles of the pulsating flow at different fractions of the period of pulsation ................................................................................................ 59

**Figure 4.28:** Frequency response curve of the combustion chamber ......................... 59

**Figure 5.1:** Coupled Helmholtz-resonators and oscillating masses connected with springs and damping elements ....................................................... 61

**Figure 5.2:** Test rig of the coupled Helmholtz-resonators ........................................ 63

**Figure 5.3:** a) Geometry of the test rig, b) definition of the flow domain, c) 3D block-structured mesh ............................................................................ 64

**Figure 5.4:** The computational domain with block structure in the symmetry plane ........................................................................................................ 65

**Figure 5.5:** Distribution of vorticity in the symmetry plane .................................... 66

**Figure 5.6:** Iso-surfaces of the $Q$-criterion at the level of $10^4$ $s^{-2}$ .............................. 67

**Figure 5.7:** Amplitude response of the coupled resonators ...................................... 69

**Figure 5.8:** Phase transfer function of the coupled resonators ................................. 69

**Figure 6.1:** Mass flow rate of the single resonator without excitation at the inlet ..... 71

**Figure 6.2:** Frequency spectrum of the outlet mass flow rate of the single resonator ................................................................................................. 71

**Figure 6.3:** Frequency spectrum of the outlet mass flow rate of the coupled resonators ............................................................................................... 73

Die VDM Verlagsservicegesellschaft sucht für wissenschaftliche Verlage abgeschlossene und herausragende

# Dissertationen, Habilitationen, Diplomarbeiten, Master Theses, Magisterarbeiten usw.

## für die kostenlose Publikation als Fachbuch.

Sie verfügen über eine Arbeit, die hohen inhaltlichen und formalen Ansprüchen genügt, und haben Interesse an einer honorarvergüteten Publikation?

Dann senden Sie bitte erste Informationen über sich und Ihre Arbeit per Email an *info@vdm-vsg.de*.

### Sie erhalten kurzfristig unser Feedback!

VDM Verlagsservicegesellschaft mbH
Dudweiler Landstr. 99
D - 66123 Saarbrücken

Telefon +49 681 3720 174
Fax +49 681 3720 1749

**www.vdm-vsg.de**

Die VDM Verlagsservicegesellschaft mbH vertritt

Printed by Books on Demand GmbH, Norderstedt / Germany